A. P Cole

Cole's Combined System of Drainage and Irrigation

.

A. P Cole

Cole's Combined System of Drainage and Irrigation

ISBN/EAN: 9783744678162

Printed in Europe, USA, Canada, Australia, Japan

Cover: Foto ©berggeist007 / pixelio.de

More available books at **www.hansebooks.com**

COLE'S COMBINED SYSTEM

– OF –

Drainage and Irrigation.

"A NEW SYSTEM OF AGRICULTURE,"

ORIGINATED BY

A. N. COLE, WELLSVILLE, ALLEGANY CO., N. Y.

BY A. P. COLE.

PRICE IN PAPER COVER, 75 Cents; CLOTH, $1.00.

CONTENTS.

❖ PREFACE. ❖

———❖———

A CONSTANT and increasing demand from every part of the world, especially from all sections of our own country, for a cheap Manual descriptive of Mr. Cole's system of Drainage and Irrigation, has at last induced him to intrust to me the work of arranging this book, he being too ill at this time to do so. In its preparation I shall endeavor to treat the subject in a most practical manner, and in the fewest possible words necessary to the end in view, and yet complete enough that all may fully understand, and for themselves see its numberless advantages and possibilities.

In the facts stated we have been conservative, feeling it better to under, rather than overdraw them.

Speaking for Mr. Cole, we extend to every reader of this Manual an invitation to visit the "Home on the Hillside," and make personal examination of his work and its results. The grounds are open to public inspection at all times.

A. P. COLE.

Biographical sketch of the life and services of the Hon. A. N. Cole, of the Home on the Hillside, Wellsville, Allegany County, N. Y.

CHAPTER I.

As a frontispiece to this volume, we open its pages by introduction of the likeness of the originator of what has been denominated aquaculture, "the new agriculture," or sub-irrigation. The relations of father and son existing between the author of this sketch and the originator of the New Agriculture, makes it a matter of delicacy to undertake this biography, and yet, feeling as we do that none know the man on whose life we discourse better than his firstborn, we waive considerations of delicacy and proceed to our work cheerfully. It would be impossible to write out in anything like detail even the more striking features of the life of a man of whom it may be aptly said he knew no youth, but entered upon the work of the full-grown man at the age of fourteen, knowing little of rest or respite for the full period of three score years, having wrought almost incessantly in one way and another for the good of his fellow-men. Inclined to leave to the pens of others, as far as may be, that manner of mention most commendatory of his work, we introduce at opening an editorial appearing a few months since in the columns of "Colman's Rural World," as follows: "Mr. Cole was born in Cattaraugus County, N. Y., in 1821, and is no longer a young man in the matter of years. He comes from grand old stock, and inherits an amount of vitality which renews its youth with passing years, hence never gets old. The Plymouth or Roger Williams Colony at Freedom, Cattaraugus County, N. Y., embraced representatives of the blue blood of nearly or quite all of the early Pilgrim families. The first great swarm that went out from Plymouth and Providence settled in the Genessee Valley and along the Cattaraugus Creek in the beginning of the present century. Others followed until nearly all the counties of Western New York, North-western Pennsylvania, and the

Western Reserve, Ohio, were embraced in the territory settled by this class of families with their sturdy New England manhood and radical characteristics. The prevailing sentiment was democratic, though the percentage in New England was quite as generally federal as democratic. A recent writer referring to this matter says, 'the Coles were fearful fighters, intermarrying with the Standish and other martial blood. The Roger Williams family and that of the Hathaways, Hoags, etc., were Quakers in spirit but would fight like devils if driven to it. This was the class of farmers who, hearing the guns at Bunker Hill and Lexington, left their oxen in the field and plows in the furrow, rushed to their barns, and seizing scythes and pitchforks, went out and took a hand in the fight. This is the stock from which our subject comes and this the blood he inherits."

That the reader may take in the situation, the author of this sketch will say that, to understand the manner of blood represented in the person of the discoverer of the New Agriculture, reference to a volume entitled "Ancient Landmarks of Plymouth," by William T. Davis, published by A. Williams & Company, Old Corner Book Store, Boston, in 1883, will disclose the fact that James Cole, of Cole's Hill, Plymouth, from whom the subject of this sketch was in direct line descended, was found dwelling near Highgate, London, in 1616, who was a great lover of plants and flowers, and married a daughter of Lobel, the celebrated botanist, who was the physician of James the First, from whom the plant lobelia derives its name. Not long after, said James Cole, having with two brothers and their families, emigrated to America, were found among earliest of settlers at Plymouth, James in possession of lands on Cole's Hill, a spot as historic as any perhaps in the Old Colony. Of the antiquity of the house in which James Cole followed the occupation of vintner, the reader may judge by the following appearing in the diary of Judge Sewall, one among earlier jurists of Old Colony fame. Judge Sewall says, at date of March 8th, 1598: "Get to Plymouth about noon, and lodge at Cole's. The house was built by Governor Winslow, and is the oldest in Plymouth." Still more notable is the name of Roger Williams, of Rhode Island, father of freedom and religious toleration in the New World, from whom in direct line came Joanna Williams Cole, mother of A. N. Cole, having been, as we understand it, a great-granddaughter of Joseph Williams, great-grandson of Roger Williams. Nor was it alone in direct lines of blood and lineage that the subject of our sketch inherited characteristics such as few men of his day and generation have seemed to possess in eminent degree. It was E. V. Smalley of the *North-west*, whose life work has so largely impressed itself on the newer states and territories, who, in a letter of recent date, addressed to Mr. Cole, says : " You put me forcibly in mind of my old friend Albert Brisbane, who held that man's chief mission on earth was to improve, beautify, and adorn the planet on which he lives." How early in life the passion developed in Mr. Cole so ruling above all others throughout, his love for plants, fruits, and flowers, we cannot say, but writing at the age of forty-four have only to say that our memory does not reach back to a time when our father's delight was anywhere so great as in home and garden. Rare fruits were his special delight, strawberries his favorite. To find out the way, and make it known to others, how to grow all varieties of plants and secure utmost profusion and perfection was his constant endeavor. That this disposition not only, but that equally of the crusader in moral, social, religious, and political progress came to a marked degree from that marriage and intermarriage among families of the Plymouth and Providence colonies discovered in genealogical lines of fully two hundred years, antedating the general exodus from these colonies into more western settlements mentioned in "Colman's Rural World," an examination of the volume referred to, entitled "Ancient Landmarks of Plymouth, will make unmistakably manifest.

When we say that the focus of moral, religious, and political propulsion so characterizing the progress of the American people, more especially in the last decade of the eighteenth, and thus far throughout the nineteenth centuries, can be no more becomingly fixed than at Freedom, Howland's Flats, Arcade, and their vicinities, where founded was the Roger Williams Colony at Cole's Settlement, Cattaraugus County, erecting churches and establishing schools in propagation of their faiths in religion and politics, we state what is simply a fact. Thence radiating outward in all directions, the influence of these settlers will be plainly discovered throughout that region where Horace Greeley was found in his early life helping his father, during his struggles in Wayne, Erie County, Pennsylvania, on the State line opposite Clymer, Chautauqua County, N. Y., while hewing out of the wilderness a home. William H. Seward was acting as land agent at Westfield, Chautauqua County, N. Y., making history as a co-worker with George W. Patterson; and where somewhat later, Joshua R. Giddings and old Ben Wade, in their humble law office at Jefferson, Ohio, were found breaking the fallow ground of freedom to all, and born and bred was James A. Garfield and others without number, making history for all time, in and outside of Ohio, now become emphatically, in the stead of Virginia, the mother of presidents. Alongside of these in Pennsylvania were found the counties of Erie, Crawford, Warren, Potter, McKean, Tioga, Bradford, Susquehanna, &c., where that young Saul of his day and generation, David Wilmot, a full head taller than that of any of his brethren, swayed sentiment to good degree, not only throughout the Keystone State, but all over the American Union. It is in this region, especially in the counties of Allegany, Cattaraugus, and Chautauqua, where in fullest force was early developed that irrepressible conflict between freedom and slavery, so for a full generation, at least, convulsing our country. It is here, whether on records of fame among the living, or on those of marble and granite, are most found the names of families of the Old Colonies of Plymouth and Providence. Here it was that A. N. Cole in early childhood, his father and mother dying when he was but about four years of age, was brought out of the woods and adopted by Asher P. and Polly Hickox, of Pike, then Allegany, since Wyoming County, N. Y., his adoption having been arranged by kinsmen. He was about five and a half years old when adopted, and during the next ten years spent not far from four years in all of study at the village school, where, such was the rapidity with which he made progress with his studies, as to become a wonder to his teachers.

Of the period at which he learned his letters he has no recollection, but infers that these were gotten at his mother's knee. Suffice it to say, that from books purchased for his reading by his foster parents, and more especially from a fine library of the times, possessed by a neighboring lady of culture and refinement, Mrs. A. A. Emery, he secured the reading of poems, a few works of historical romance, volumes treating on navigation and discovery, &c., &c. Among all of these books none made a deeper impression than a little volume entitled the "Life of Benjamin Franklin," written by himself. Here he became acquainted with "Poor Richard," so like to Horace Greeley of later times, and from "Poor Richard's Almanac" the lad gleaned an amount of knowledge impossible to calculate. He had reached the age of eleven, and somewhat upward, when somewhere finding, we think, in this renowned calendar, a mention of Chinese agriculture, he was prepared to learn with avidity more upon the subject as found in one of the most remarkable books published up to that period, to wit: the "Uni-

versal Geography," of Jedidia Morse, which was possibly the most compre-
hensive and complete summary of geographical and historical events embodied in any volume of the times anywhere. Obtaining the book, under
the head of "Chinese Agriculture," page 417, second volume, he found the
following : " The Chinese agriculture is carried to a high state of improvement. On the sides of their steepest hills terraces are formed, supported
by walls of stone, and the whole mountain is cultivated to the summit,
where reservoirs are sunk, into which rain waters are collected and conveyed
around the terraces down to the base of the mountain. With great toil the
people collect every species of manure. Prodigious numbers of old men,
women, and children are constantly employed in the streets and on the
banks of the rivers and canals, with baskets tied before them, collecting
every kind of manure. The Chinese farmers apply liquid manure to the
roots of their plants and trees. They also steep their seeds in liquid
manure until they germinate, before they are sown."

This item of knowledge, touching the methods in agriculture practiced
by a people whose progress as regards certain arts and sciences antedates
those of perhaps any other nation, opened up a vista of the future to the
eyes of the boy reading it, having seemingly the bringing out of a system,
which. quoting from a leading editorial of the late and lamented William
Dorsheimer of the New York *Star*, appearing February 11th, 1887, under
head of

No More Drought.

It will be remembered that some months ago the *Star* opened its
columns to the discussion of a most interesting theory of the possible preservation of water. It was Mr. A. N. Cole of Wellsville, Alleghany County,
N. Y., who first propounded and advanced the proposition that it was willful
waste to allow the rains and snows to form into streamlets, rush into rivers,
and bear out to the oceans the richest portion of the soil, often leaving ruin
in their course. He asserted that by a simple adaptation of the lands adjacent to the natural watersheds of the world, reservoirs of all sizes could
readily be obtained, in which the waters would accumulate and be conserved,
and from which, by an inexpensive system of sub-surface irrigation and
drainage, they might be released to moisten and feed the dry and hungry
earth.

Mr. Cole was the life-long friend and associate of Horace Greeley.
Together they worked out many problems for the amelioration of their
fellow men.

Mr. Cole alone has survived to witness the fulfillment of a dream that promises to revolutionize all agriculture, horticulture, forest culture, fish culture, manufacturing, mining, and inland navigation.

On his hillside farm at Wellsville he has already put his theory into practice, so far as crops are concerned, and demonstrated the feasibility of reserving a surplus of water from the wet seasons for use in times of drought. He has lived to see the subject of aquaculture taken up, discussed, and advocated by the leading agricultural journals of the day. The *Rural World*, published by United States Commissioner of Agriculture, Norman J. Colman, has this to say of the old philosopher's wisdom and ingenuity:

"There are few sections of this vast country where water is not furnished naturally by means of the rains or the snow, but their coming is comparatively uncertain, and very often when most wanted, nay, when absolutely essential to the production of a crop, they are not forthcoming. Did the Creator ever intend that because he sends a surplus now it should be permitted to go to waste simply because it is not for the moment wanted?"

We reprinted yesterday a most interesting article from the *American Angler* which presents this great subject in all its most attractive lights to farmers and kitchen gardeners.

We are aware that Mr. Cole has been associated with vast enterprises, is a pioneer in great undertakings. We remember that he assisted in putting the Republican party on its legs; that he outlined our Central Park and foreshadowed our systems of rapid transit; but of all the work that a keen mind, an unflagging industry, and an unfaltering faith can accomplish, we submit that in thus understanding and defining the economies of nature's water supply, he has proved himself to be one of the greatest benefactors of the age.

As, amid countless achievements, the one of the New Agriculture is that bound to live after him for all time, and the one to which this volume is devoted, no sketch of his life would be other than the play of Hamlet with the part of Hamlet omitted, which failed to make mention of the step by step leading him to ultimate discovery and demonstration of his methods. Before proceeding further, we will here say that, perhaps had the entire world been gone over, no child of his age could have been found by the time the subject of this sketch had reached the age of twelve, who had looked more deeply down into that cycle of the waters embracing the mysteries of evaporation, the formation of clouds, condensing of vapors into

rains and dews, their congelation at certain altitudes, precipitation of rainfalls, the falling of snows, with like causes and effects, inclusive of the origin
of springs, and all that relates to the movement of waters on, along, and
beneath soils everywhere the world over.

Such was his rapt attention given to looking into the secrets of nature,
and such his constant reading of books, as to make him, whether at school,
at home, or among the people of the town, an object of interest. His foster
parents were devotedly attached to him, and did all in their power to aid
and encourage him in the acquisition of knowledge.

In nothing, however, did he seem in childhood to take a deeper interest
than in the one of parties, politics, public men, measures, etc. Educated ·
after the strictest sect a Democrat, or, more properly speaking, a Jeffersonian
and Jacksonian Republican, to propagate that sort of faith and shun federal-
ism as a heresy not to be tolerated, became early a passion. Reading the
Albany *Argus* and *Evening Post* weekly, the teachings and writings of
William Cullen Bryant, William Leggett, and others of more advanced
ideas of the times, made him fundamentally acquainted at an early age with
the sort of democracy found in the Declaration of Independence, and form-
ing, in fact, the foundation of the Republic. As for Federalists, Free Traders,
and born nullifiers, such as John C. Calhoun, Aaron Burr, and their kith and
kin, the boy at twelve was as antagonistic as the man of the present, nearing
that proverbial period of life allotted to man. So decided, indeed, has been
his democratic bias all along through life, and thus remains, he has never
been able to become aught else than a defiant fighter against the sort of
politics which calls for subordination of individual conscience to dictates of
caucus, cabal, or of star chamber emanations. Of his inborn and impulsive
radicalism, the reader may judge by a single incident occurring at the age
of thirteen, when, entering the village mill, and discovering a paper pinned
upon the wall with a few names appended to it, all of men of maturest years,
he read it, and finding it a pledge to abstain from the use of alcoholic drinks,
he at once added his name to the list, and hence his claim to being one of
the earliest among temperance men now living is well founded. Only a
year later, the foster father's health having given out, his business, that of
keeping a modest country store, had to be abandoned, and at about the age
of fifteen the lad bade adieu to school with its lessons, and to the best of his
ability set himself at work to settle up the business. To conclude that, at
this time, young Cole had mastered as much of science as perhaps any one
of his age in the county of Allegany, and was possessed of as large a fund

of knowledge as any in the entire State, is at least a reasonable presumption, since he had mastered Daboll's arithmetic, Woodbridge's geography, Murray's grammar, all of the common branches of the times, as they were denominated, had done an amount of reading impossible to estimate, and on quitting school, outside of his passion for trout fishing and making experiments in growing plants, giving necessary attention to the settlement of his foster father's business, knew no rest, but pushed on all the more vigorously with his reading and study of books, making good beginning with algebra, chemistry, botany, mineralogy, natural philosophy, geology, etc., with the elements of which he became more or less familiar, and had made some progress in Latin, when, having reached the age of sixteen, his foster mother, finding herself under the necessity of breaking up housekeeping, he contracted to teach a winter school in one of the most benighted districts of Allegany County, comprising territory in the townships of Hume and Canadea, in which was found settled a few estimable families, but the bulk of the district made up of what were denominated at times "the lost nation," and quite as commonly known as "swamp angels." There were between twenty and thirty scholars in the school, their ages varying between ten and twenty; the older ones, as a rule, being little less than barbarians. It only required five weeks of endeavor to teach such a school before the big boys and girls of the clay-bed district were masters of the situation, and the boy schoolmaster found himself as frequently out as inside of the school-house. Reluctantly, and yet from necessity, the stripling surrendered, and, returning to the village of Hume, joined his foster-mother, the two securing board at fifty cents per week each. The late teacher himself again became a scholar in the village school, and at the close of that winter's term was sufficiently advanced to give lessons not only to the hundred pupils attending the school, but equally to his teacher, a roving Texan ranger, who, by chance, coming that way, had taken the village school, which he succeeded in teaching to conclusion, amid as much of fight and fury as possible to imagine, the peace only kept, in the meantime, by surrendering the more advanced classes to the tutorship of young Cole.

The early spring succeeding, while snow was yet on the ground unmelted, the young man, having decided to try his fortunes in the West, left Hume for Bloomfield, Ontario County, to visit friends before taking a packet on the Erie Canal for Buffalo at opening of lake navigation. It was the last of April when he crossed the gang-plank of the "Cleveland," at that time one of the chiefest among lake steamers, and the first of May found the young man in that city of Ohio bearing the name of the ship that carried him safely. As he stepped upon the dock he found himself with barely money sufficient to last three or four days by closest economy, and at once sought for some sort of work to do which would give a living, however scanty, and this he found by contracting to keep in order the editorial rooms of the *Cleveland Intelligencer*, a daily and weekly, published by Benjamin F. Andrews, postmaster of a town at that time containing only about six thousand population, with an area, inclusive of suburbs, so extended that to estimate the ground covered by the young city would make it incredible with every reader when we add that, to his duties in keeping the editor's rooms in order, were added those of carrier to the doors of city subscribers. His compensation for this work was merely nominal, and by the first of June, the cholera having broken out, the young man, exhausted with overwork and no pay—in fact, having been offered an agency to canvass for a modest little monthly, the *Buckeye Ploughboy*, subscription price fifty cents a year, with commission of twenty-five per cent., accepted with alacrity, and, from the hour of doing so, set out on that race which has been one of undeviating progress for a period of a full half century, having few parallels in American life. This has not been in the accumulation of wealth, since he has seen much of stringency in money matters—much of embarrassment, in fact; and, though he has made, possibly, more money during that fifty years which has been absorbed by the rich and powerful, profiting by his work, than any other American, he has himself, up to this time, comparatively little to show of earthly possessions, and yet has lived all the while since the autumn succeeding that never-to-be-forgotten summer of 1838 in comfort, blessed with goods which come of an abundance of all required for the wants of man.

Many a time has the author of this sketch listened to the accounts given by his father of his experience, traveling on foot hundreds, not to say thousands, of miles over the Buckeye State, adding several hundred names to the subscription list of one of the earliest among agricultural and horticultural journals of the then West, since Ohio was reckoned in those days central among States of the West. Since the settlement of our country no season has proved more disastrous to crops than the one in Ohio of 1838, coming of a drought so unparalleled that, at time of potato-digging in October, the tubers only averaged, as a rule, from the size of a walnut to that of the butternut, nor were of value outside of seed for the ensuing year.

Fruit was abundant, more especially peaches, which rotted upon the ground all over the State. Wheat matured early, and proved a crop from fair to middling; while corn was, as a rule, a failure. So early as July first, such was the universality and severity of the drought, that the springs nearly all disappeared; the creeks, brooks, and rivulets became dry in their beds, and great rivers shrank to comparative nothingness in volume. It was said of the Sandusky River that it was to be discovered only in detached pools· and so thick did the scum form on the surface of these as to enable the squirrels to cross them in droves, as upon a bridge. There were but few wells in the State at that period, and these were generally dry from July till November. Cattle and sheep died by thousands for want of water; the swamps were made little else than burning bogs of peat and tinder; nearly all of their timber destroyed by fire. It was shinplaster days; no bank bills worth the mention; and no metallic currency at all in circulation. The only subscriber to the *Ploughboy* paying for it in silver was none other than the Hon. Joshua R. Giddings, who, alongside of Ben. Wade, was found in their law office at Jefferson, Ashtabula County; and here began an acquaintance proving of incalculable value to the then still stripling—now a veteran crowned with locks as white as snow. "Whence come you, my boy?" asked Giddings. "From Western New York, Allegany County," was the canvasser's reply. "Of New England parentage, I venture to say?" continued his interlocutor. At this the young man, never averse to speaking of his ancestry, spoke of the Coles, of Cole's Hill, Plymouth, and pointed back to his mother as a descendant in direct line from Roger Williams. This was sufficient. It was Saturday afternoon, and the "Old War-horse of the Western Reserve" invited the canvasser to spend Sunday with him. This invitation was joyously accepted, and when, on Monday morning, he made his departure, it was with a head full of new ideas, and a heart more than ever before upwelling with love of liberty. It was only June at this time, and, before the season was over, the young man at times made his bed in the fields, alongside of the shocks of wheat and corn; and, while making his evening and morning meal of fruit and raw wheat, shelled out and eaten with his then young and firm teeth, he was put impressively in mind of the brief stay made at the house of Giddings, who, later in 1853, came down to Western New York and joined his early disciple in pushing that crusade for freedom and equal rights resulting in the greatest moral, social, religious, and political revolution in the annals of time. Only a few days after leaving the home of Giddings the fearful drought began of which we have made mention. It was during its prevalence that the lad began looking more deeply than ever into the **origin of springs,** rivulets, rivers, and lakes, **and in contemplating the** devastations wrought by the drought, **began mentally inquiring whether it was not possible to save and store the rains of spring, summer, and autumn, as well as the melting snows** of winter—doing, in fact, what he has done and demonstrated since by his methods **of aquaculture, the new agriculture, or sub-irrigation—** more properly speaking, *subsoil* irrigation. That at eighteen years of age he had gotten more than an inkling of his after-discoveries **well fixed in** his mind, let no reader **for a moment** doubt, since that mention of **Chinese** agriculture found in a geography, published several years before he was **born,** was a something borne constantly in mind and studied upon, **and ·** **had** opened to him vistas of a future at last reaching out over all of the face of the earth.

The close of September found the young man **at** the home of **an un- cle,** Nathan Cole, in Republic, Seneca County, **and** here it was arranged for a winter school, the teaching of which for four months, at twelve dol- **lars per month, laid the foundation of future** and undeviating success. **His patrons and pupils were nearly all** Germans, " Pennsylvania **Dutch," as denominated, and few could speak, and none among them worth the** mention **could read,** English at all. It was, in fact, the first English school **taught in the district, and the young teacher left his charge at end** of his term **with the good-will and followed by the blessings of parents** and children, making **his way to Michigan, at that time just emerging** from the wilderness and **putting on** the vestures of Statehood. **A severe** attack of bilious fever, succeeded by ague and fever, followed immediately **after** reaching Michigan, and necessitated earliest possible resumption of **work,** when the first of June, 1839, found the young man teaching a sum- **mer school** at Northville, Plymouth County, at two dollars per week, **boarding round.** On reaching Northville the first object specially attract- **ing his attention** was that phenomenal flood of pure spring water, putting **him in mind of the** springs at Caledonia, Livingston County, N. Y. **On** these wonderful springs are at present **located the** principal hatching **and** breeding ponds **of the** United States Government, and it was this remark- able outgushing **of pure** spring water **at sources of** the Rouge River, that gave to the place **the repute of being the healthiest in** Michigan, attract- ing to it parties seeking **to avoid malarial influences so generally** prevail- ing in most portions **of the West of those days.** Whence came these great fountains? was the first question **arising in the** mind of **him who** seemed so born to the following after the waters that the passion would somehow never down. The young man was not long in making up his mind whence came, and from what conditions sprang, the wonderful springs. Less than a month after beginning his school, among others seeking to find a healthful spot in which to spend the summer, was M. Theodore Lupien, a young Frenchman of eminent family, accomplished in manner, and having much of learning acquired in best of schools. Only a few words of English could he speak, and, placing himself under the tutorship of the young teacher, with a proposal to make a Frenchman of Cole, while Cole in turn was to make an American of Lupien, as regarded language, at least. The work of education went on. We had omitted to mention the fact that in teaching among the Germans of Ohio, the preceding winter, the schoolmaster had gathered the fundamentals of the language of the Fatherland, and this, added to rudiments of Latin, helped him on with the French, and a month had scarcely passed before the teacher of that summer school, then about eighteen years of age, would have proven a creditable professor of the French language in most American schools of the times. Not so with his pupil, however, who made but slow and tedious progress in learning to read and speak English.

At close of his four months' term of school A. N. Cole left Michigan for return to the land of his birth, reaching Allegany and Wyoming counties in the month of September, 1839. Professor Davis W. Smith, of Castile, Wyoming County, having acquired an enviable reputation as a teacher, young Cole at once availed himself of the opportunity of study at his school, and though there was in the school a considerable number of scholars well advanced in learning, years older, in most instances, than Cole, the latter, as usual, took the lead, attracting general attention as a student of rare promise. Among others attending the school were a son and daughter of Professor Joseph Wildman, principal of the first High School at Poughkeepsie. The family of Professor Wildman was one to which the subject of this sketch found early attraction, and here it was that, a few months later, the young man found the home to which his heart to the present hour turns back as one of love and beauty having few equals anywhere. Here, at the age of two-and-twenty, he was wedded to Margaret Malvina Wildman, and never was union more perfect, and never did the lives of any couple move more even and happily on. The fruit of this union was four children, three sons and a daughter, still living, though their mother left this world for Paradise, March 22d, 1880.

It was Joshua R. Giddings who, in the summer of 1838, suggested to the subject of this sketch the name of Horace Greeley as the one thus early giving evidence of becoming a second Dr. Ben. Franklin, and advised the lad to form his acquaintance. This advice was taken, and Cole wrote his first letters to Greeley in 1839 or '40, while the latter was publishing the *New Yorker*. These letters were cordially received by the now everywhere admitted greatest among journalists of any age or country. The letters dwelt mainly on subjects of social and political reforms, the theories of "Fourier," "Brisbane," and others of their ways of thinking. In the meantime, both Greeley and Cole had become thorougly imbued with the teachings of M. Francois, Pierre Guillaume Guizot, whose philosophies were reflected by his renowned society, claimed by many to have a beginning so early in the ages to make it impossible to trace its origin, or fix the date of its first appearance, and yet here is what has been said of it as matter of history. The name of the society was "*Aide toi et le Ciel T'aidera*," signifying, "Heaven will help those who help themselves."

This was the motto of a political society having for its object the attainment of ends best defined by a common platform now occupied in substance by the Grange, as well as by the farmers, foresters, fish culturists, and other industrial classes in alliance, seeking such reforms in governments everywhere as embodied in the following general declaration of principles:

Profoundly impressed that we, the farmers of America, who are united by the strong and faithful ties of financial and home interests, should, when organized into an association, set forth our declarations of intentions; we therefore resolve,

1. To labor for the education of the agricultural classes in the science of economic government in a strictly non-partisan spirit, and to bring about a more perfect union of said classes.

2. That we demand equal rights to all and special favors to none.

3. That we return to the old principle of letting the office seek the man, instead of the man seeking the office.

4. To indorse the motto, "In things essential unity, and in all things charity."

5. To develop a better state mentally, morally, socially and financially.

6. To create a better understanding for sustaining our civil officers in maintaining law and order.

7. To constantly strive to secure entire harmony and good will to all mankind, and brotherly love among ourselves.

8. To suppress personal, local, sectional, and national prejudices, all unhealthful rivalry, and selfish ambition.

9. The brightest jewels which it garners are the tears of widows and orphans, and its imperative commands are to visit the homes where lacerated hearts are bleeding; to assuage the sufferings of a brother or sister; bury the dead; care for the widows and educate the orphans; to exercise charity toward offenders; to construe words and deeds in their most favorable light; granting honesty of purpose and good intentions to others, and to protect the principles of the National Farmers' Alliance and Co-operative Union until death. Its laws are reason and equity; its cardinal doctrines inspire purity of thought and life; its intention is, "Peace on earth, good will toward men."

At the opening of this sketch occurs the introductory on the part of Colman's *Rural World*, which we now supplement by the following from the same source:

Mr. Cole commenced life in the second quarter of the present century, a time which is remembered as producing many of the greatest men that have figured in the history of this country, and he has been of them a part, their daily associate and correspondent, and a co-worker with them in making history and the nation's greatness. With a memory of extraordinary retentiveness, and a more than ordinarily active and vigorous mind, he naturally retains all of the past, and when the occasion requires, pours it forth in a volley and volume which, for force and fervor, is seldom equalled. He reminds one, indeed, of his old and life-long friend, Horace Greeley, for neither his tongue or pen are capable of idleness, not even when asleep; indeed, we never met so prolific a writer or one having greater facility of expression.

"This, then, is the author of the 'The New Agriculture, or The Waters Led Captive,' by whose genius and indomitable perseverance we believe the desert is to be made to blossom as the rose, the fruit and vegetable gardens to produce three or four times the quantity and two or three times the size heretofore known, and the waters to be so saved and held as that drouths shall be unknown as long as rains fall or rivers run. We have heretofore expressed the opinion that there was no more sense in permitting a waste of waters when they came as one of nature's largest and best gifts to man, than in allowing the crop of corn or wheat, of fruit or vegetables, to rot where they grew. We are of that opinion still. And whilst we read of the many advances in science and scientific commerce, in art and literature, in education and morals of the Victorean era; whilst we enumerate the wonders of steam, of electricity, the printing press, and of agricultural and commercial progress of the past fifty years, we look to the remaining years of the century to perfect this still more magnificent, this grandest of all enterprises, the new agriculture, or the waters held captive.

" Six years ago Mr. Cole's neighbors were astonished by the growth of fruits and vegetables of marvelous size, beauty, profusion and perfection on the heretofore barren and rocky hillside of his suburban home. He was understood at first as making experiments in underdrainage ; nor did his nearest neighbors and most intimate friends have any intelligent conception of the methods under which he was proceeding—those of the sub-surface, subterranean or underground irrigation— but he brought that rocky hillside into such a state of fertility that finer fruits and vegetables than were ever seen, and in greater profusion, were produced therefrom ; then and not till then did he patent the system, and give to the world the grandest idea that man in agriculture ever conceived. In thus writing we are conscious of the fact that many otherwise excellent men and able writers have both ridiculed the system and villified the man, but he will live to see them take it all back and acknowledge that he is not only right, but the benefactor of his age and people."

As politics and parties should have nothing in fact to do, as such, with ways and means by which the industrial classes are to find their way up and out into conditions of universal thrift and prosperity, what we have said touching the society of Guizot should be accepted therefore as best of reasons why in a political sense no further mention be made of Mr. Cole's identification therewith.

This book is devoted to the New Agriculture, the fundamentals of which are best bought out in a letter of Dr. John Swinburn found on its pages. This is the water feature whereby towns, cities, all portions of the world, indeed, are to be supplied with an abundance of pure spring water. It is the Croton water-shed pointed out in conclusion by the pen of General Charles A. Dana of the New York *Sun*, appearing at date of November 14th, 1888.

THE GREATEST OPPORTUNITY.

Now is the time for the Vanderbilts, or any other set of enlightened millionaires, to come forward and undertake here in this neighborhood an experiment whose successful working would confer upon the human family a greater benefit than any novelty or invention or discovery since the introduction of printing.

We refer, of course, to the New Agriculture, the great system of subterraneous irrigation, of feeding the roots of plants from beneath with a perpetual supply of moisture. This system was discovered by that irrepressible, electrical veteran, Asahel Nichols Cole of Allegany County, and the right place to make a conspicuous and triumphant display of its marvellous results is here at the doors of this metropolis, among the hills of Westchester. The land is there, its long slopes turning to the southern sun; the living springs of water are there; the climate is favorable, the situation peerless, and all that is necessary is that some great and far-seeing man, with as much money as he has brains, should devote a little thereof to a work whose success will not merely make its capitalist glorious and famous, but also increase his wealth beyond the wildest dreams of avarice. No matter how many millions he may have already, the New Agriculture would add to his store, and, in addition, the blessings of the human family, the cry of joy from poverty relieved, the shout of hope from hearts that dread and doubt, would be given to him in full measure and exulting chorus.

What wise millionaire, what rich and great philanthropist, desirous of being the benefactor of the human race, of putting an end to hunger and poverty, will come forward and lay hold of this unexampled opportunity to gain for himself imperishable renown, and to confer upon his grateful countrymen the benefits of universal prosperity and boundless abundance?

CHAPTER II.

Water a Necessity for Vegetation.

Years ago, when a large portion of the country was covered with forest, falling leaves, decaying woods and other vegetation protected the soil making it porous and light, with power to absorb and hold the rain-fall almost in its entirety. Floods and freshets were prevented, springs and streams gave a full and steady flow, while the rain-fall was more abundant and regular. Soil then, when cleared for crops, filled with decaying remains of the removed woods, readily took up and held for a longer time the waters which fell upon it, and drew also upon the surplus moisture contained in the forest lands adjoining or surrounding it. Then crops had a constant supply of water, and yield was far in excess of later years, and of superior quality. The forests have been removed; their rain-absorbing soil has given place in large part to pasture and meadow lands, with their compact surface and almost water proof mat of roots. These lands have the power and capacity to absorb but a very small percentage of the falling rain, and it is lost to crop and soil in wash and flood. Springs have failed; streams have lost their volume in large part if not wholly during the heat of summer; evaporation from them being shut off, and that from the soil increased. The annual rain-fall has steadily decreased, drying winds circulate more freely and largely, drouth becomes more frequent, severe, and extended. Every farmer in the majority of years feels that some at least of his crops would be greatly benefited by an increased supply of water during the growing season.

Market gardeners and fruit growers, whose crops should on the average reach in value several hundred dollars per acre annually, experience large losses which might be avoided were a supply of water at all times available. Not only is the average yield of crops growing less yearly, even in favorable seasons, but drouth is becoming so general that crops suffer and are in whole, or large part, destroyed about once in three years.

Every tiller of the soil knows that water is indispensable to the growth of his crops, and if it is not supplied, the crop correspondingly suffers.

There are some facts, however, touching vegetable growth which are not generally known, but should be understood by every cultivator of lands. Henry Stewart in his book, "Irrigation for Farm, Garden and Orchard," states them in the following plain and direct manner. "No water, whether it be in the state of liquid or vapor, can enter into any part of a plant other than its roots. The common idea that water or watery vapor is ever absorbed through the leaves of a plant is unfounded. The solid portion of the plant consists of matter which enters into it only while in solution in water. Water is the vehicle by which the solid part of a plant is carried into its circulation for assimilation. If water is not adequately supplied, an insufficient quantity of nutriment only will be carried into the circulation of the plant, and its growth will be stunted or arrested altogether.

Growing plants contain from 70 to 95 per cent. of water. To the extent that water supplies this necessary constituent of a growing plant, it is an actual nutriment.

The quantity of water that must pass through the roots of a plant of our ordinary farm crops, and to be transpired through the leaves, to carry it from germination to maturity, is equal to the depth of twelve inches over the whole soil covered by the crop. This is the requirements of an average crop upon a moderately well-cultivated soil. If the crop is stimulated to extraordinary growth by large applications of manure or other fertilizers, a still greater supply of water is needed to meet the demands of the crop."

These facts so plainly stated show the most potent factor in the production of crops to be soil moisture. The presence of water renders plant growth possible, as under the proper conditions it renders soil fertility available to the plant and insures full development in growth. Under old and present methods of farming and gardening this moisture is not supplied. The rains of fall, winter and early spring come at times when crops derive but little if any benefit from them. The rain-falls which occur between seed time and harvest are not only generally short of the required amount, but the evaporation of the growing season exceeds the rain-fall, and after having absorbed that, takes from soil and crop the little moisture left from the rains of winter and spring. This being exhausted, crops not only thirst but starve.

So important has become this matter of soil moisture, and so inadequate the supply for the summer months, that the practical farmers of this State who are in charge of the State Agricultural Experiment Station, declare in their Report of 1887, that "to control the water of the soil, even to some extent, is to the farmer a matter of great consequence." Agricultural papers, writers, and speakers are agitating the subject and urging a change of present methods. Most emphatic of these utterances are those of Colonel H. W. Wilson, in a recent address before the Massachusetts Horticultural Society, as follows: "About 50,000 gallons of water are ordinarily required to give an acre of land a proper saturation, and no irrigation can be at all satisfactory which attempts to do any less. As the gardener has often observed, both in the green house and in the garden, a slight watering often proves only an aggravation, and benefit is derived from a thorough drenching; so in our climate, with ordinary soils such as are found to be advantageously cultivated, it will require about two inches in depth, over the entire surface, to make a useful irrigation of almost any crop. This, with what will be lost by leakage and evaporation, will amount to 50,000 gallons. For vegetables and small fruits the value of water would be greatly increased in dry years, while for strawberries the benefit would be greater than anything of which cultivators have hitherto dreamed. Drouth is the constant dread of the strawberry grower, as the strawberry is a thirsty plant and seldom gets water enough."

Irrigated lands are lands that always produce large crops, regardless of the seasons or rains. It is a mistake to suppose that irrigation is only suitable or profitable in regions devoid of rain. While an absolute necessity in arid regions, it is a great help to successful agriculture in every section of the earth. The farmer who resorts to irrigation and drainage is always certain of a large yield; he loses no seed, labor, or crop by dry or wet seasons. Crops grown under these circumstances are always harvested in good condition and are of superior quality, consequently bring higher prices.

The following, from the *Drainage and Farm Journal*, published at Indianapolis, Ind., by J. J. W. Billingsley, is so pertinent to the question of the necessity for an ample water supply we make quotation therefrom :

"There was little call for considering the need of irrigation when the soil was fresh from the covering of the leafy mold. Then it readily retained moisture. The greater difficulty was to free the soil from excesses of moisture. Subsequent cultivation has changed the mechanical condition of the soil, and the removal of the forest timber has exposed the surface to the drying winds, until drouths are more frequent and the effects are oft-times disastrous to the growing crops.

Farmers have been casting about for a remedy, and it has been found largely in the underdrainage of the soil to a sufficient depth to remove from the soil spaces the excess of water, retaining only the water of moisture, promoting the capillary flow of moisture to the surface from greater depths in the soil, and the condensation of moisture from the atmosphere in its circulation through the soil and subsoil. But the husbandman finds that there is still a greater amount of moisture needed to promote the growth and largest productiveness of some crops than is supplied by underdrainage, or tile drainage, as it is commonly called; which is especially true in market gardening, where the value of the crop often aggregates, or should aggregate, several hundred dollars per acre. To fall below a remunerative yield is disastrous. The soil may be underdrained, manured and brought to the desired fineness in the division of the particles, but the extra supply of water must be provided for the extra yield required to make the business pay.

To do this, many devices have been brought into use to provide the needed extra supply of water. Trenches or canals have been dug miles in length to convey the water to the point where it is needed, then lateral surface trenches convey it to the area of land to be irrigated, spreading it out into smaller and still smaller temporary trenches, until the supply is made as uniform as the amount of water, time, and labor required will allow.

Wells have been dug and the water pumped into trenches to be distributed to various parts of the ground. Engines have done the work of pumping the water into tanks built upon elevated frame-work, and the water drawn off to irrigate the soil by means of pipes, hose, lead troughs, etc. Others have tried the sprinklers drawn by horse-power, and men have tried the hand sprinkler, but it is expensive, and most frequently not up to the measure of effectiveness desired.

How frequently it occurs that the labor of years may depend for its desired remuneration upon the supply of moisture that is furnished to growing and maturing crops in the short period of ten days. If the supply of water from rainfalls is cut off for that length of time when the crop is maturing the loss may prove ruinous. The question of the hour with the market gardener and small fruit grower is, how shall this supply be provided for certainly?"

This question of a uniform and adequate water supply is not alone a question of importance to the individual land-owner, but also of State and Nation, and as such is receiving at the hands of the leading men and public journals of the day the consideration its importance entitles it to. In the N. Y. *Tribune*, date June 23d, appeared the following:

"Proposed Mountain Reservoirs.

The rapidity with which in this age of occupation and excitement even the deepest impressions are effaced is strikingly exemplified in the fact that though scores of bodies are still mouldering beneath the unexplored ruins of Johnstown, the great disaster has faded out of public attention. While the horror was still fresh there was an eager desire to learn if other dams were ready to give way before an extra pressure, and the demand for investigation and prompt precautions against similar catastrophes was universal. But already the search and the discussion have been practically abandoned.

These observations are suggested by an article in the current number of *Garden and Forest*, upon the vast scheme of mountain reservoirs proposed by the Director of the Geological Survey as a means of irrigating the dry plains beneath them. There is little doubt that engineering skill is competent to carry out the plans of Major Powell, and that a great territory can be rendered fruitful through the accumulation and distribution of water which otherwise would be largely wasted; but there are serious objections which ought at least to be considered far more attentively than they yet have been before the Government goes further in the direction of committing itself to this project. As *Garden and Forest* suggests, there is something to be said in favor of the decrees and processes of Nature as opposed to the impatient purposes of man, but we are not desirous now to dispute the immediate utility of the plan for making the desert blossom. The question of safety is of the first importance. As *The Tribune* remarked directly after the Conemaugh disaster, and as *Garden and Forest* now urges, the chief element of danger in an elevated lake artificially formed lies in the difficulty of maintenance, not of original construction. Secret and obscure agents of destruction are constantly endeavoring to undermine the strongest fabric, and the obstacles to an adequate determination of their success or failure are practically insuperable. But aside from this constant menace, sudden and irresistible forces may be developed at any moment. Our contemporary points out that an earthquake so slight as to be otherwise harmless might make a breach in one of these dams through which the sea of water behind it would pour out in an instant, and it furthermore supplies this most suggestive illustration of what might follow the proposed cutting away of the mountain forests: 'The Ardeche is a small mountain stream in France, and yet the sudden melting of the snows in the deep valleys at its sources so swelled its current that it once delivered 1,305,000,000 cubic yards of water into the Rhone in three days. For this short period it flowed with a volume like the Nile, and what reservoir could be trusted to restrain an outpouring of this sort?'

These are considerations which ought not to be lightly dismissed. We are far from failing to appreciate the beneficent transformation which might reasonably be expected to result from an artificial distribution of moisture on the arid land which Major Powell designs to fructify, but we do insist that several momentous arguments against his plan ought to be satisfactorily answered before the first step is taken toward putting it into practical operation."

An even distribution of moisture over the arid lands would not only cause them to blossom, but would by even and extended evaporation cause cloud formation and rain-fall. In this book will be described a system which will receive, store and distribute the waters of the mountain streams fed by melting snows of mountains, snow-clad during a part or the whole of the year. Not only will it store the waters unattended by the dangers of the plan spoken of in the above quoted article, but it has the advantage of immediate distribution, and that beneath the surface where it will do the greatest amount of good, and be much less subject to rapid evaporation.

CHAPTER III.

Drainage.

If with a system of irrigation a proper system of drainage be also combined, the tiller of the soil will remove two adverse influences against which he now contends. Water is a good servant but a bad master. To make it our servant, not only must be it supplied in sufficient quantities, but it must be supplied in such a way as to perfectly control its effects upon soil and crop.

Stagnant water upon or near the soil's surface, works injury in many ways. It prevents early working and seeding, remains cold and becomes sour, generally destroying the seed of crops put upon the ground, or drowns out and rots the root of any which may chance to start. With the coming of hot weather, baking of the surface follows, cracks and fissures in the soil admit the burning rays of the sun and the heated air, and partial, if not total, destruction of the crop is the result.

Drain, and the stagnant water is carried off, the texture of the soil improved by being made more porous, drier, looser, and more friable and light ; air and the rain fall are more perfectly absorbed, and the soil made ready for tillage some weeks earlier by soil heat. Still farther we quote from the *Drainage and Farm Journal.*

"The amount of heat in the soil is a matter of the highest importance to the farmer. The reader need not be told that it has a great influence upon the germination of seeds and the growth of plants. A certain amount of heat in the soil is essential to germination and growth. A certain amount of heat is also essential to the decomposition of carbonic acid by plants. Indian corn will not decompose carbonic acid at a temperature lower than about fifty-nine degrees. If the temperature of the soil can be raised, plants will grow faster. Our farm crops may wither in midsummer; but this is not because the soil is too warm, but because there is a lack of moisture in it, or, what is oftener the cause, the atmosphere is so dry that the moisture is exhaled too rapidly from the plant. It would be very advantageous to increase the

warmth of the soil in the spring, for by so doing we could get plants
started several days earlier, and their growth would be more rapid. In
the North, especially, the seasons are too short for many plants; and
even those which mature well enough would make a heavier yield if
started earlier and forced along for the first few weeks by a warmer soil.

Drained soils are warmer than undrained ones, because much of the
water which passes through them would otherwise be evaporated; and
this evaporation, it must be remembered, requires enough heat to make
the water into vapor. The amount of heat thus used is quite large. It
is likely equal to that produced by burning two-thirds of a ton of coal per
day for each acre on an average throughout the year. If this heat were
not absorbed by the water, it would be largely absorbed by the soil, if
all of it were not. The specific heat of water is greater than that of the
soil—five times as great as humus, seven times as great as loam, eight
times as great as clay, and ten times as great as sand. Hence, more
heat is required to warm up a certain weight of water than to raise
the same weight of soil to the same temperature. As drains remove an
excess of water in the soil, in the spring at least, it would remove water
that otherwise would take heat from the earth. More than this, water is
a very poor conductor of heat, and when the soil is wet, as in the spring,
the heat will penetrate much more slowly than if the water were removed.
In Prussia it has been officially determined that on an average the snow
there melts a week earlier on drained than on undrained land; and the
difference would not be less in this country. A difference of ten to fifteen degrees in the temperature of drained and undrained soils has frequently been noticed ; and the constantly higher temperature of drained
soils is doubtless responsible for much of the larger growths upon them."

Many good points have and can be given in favor of under-drainage.
Of these the following three, by Professor R. T. Brown, are among the
strongest we have seen :

"1st. A clay soil finely pulverized and throughly dried will hold
from 80 to 100 per cent. of its weight in water before the same will drip
from it. This water of sub-saturation, known in common language as
the moisture of the soil, cannot be reduced by drainage, it can be removed only by evaporation. But at this point the water is adhering to
the particles of clay, and the interspaces between the particles of clay are,
or should be, filled with air. This is a condition of the first importance
in regard to the fertility of the soil. But in a state of saturation these

interspaces are filled with water, and of course the air is excluded. Now, as this water of saturation can penetrate the subsoil but slowly, if at all, it must be removed by evaporation from the surface. But as the surface dries, the water from below rises by capillary attraction to fill the place of that removed, and this leaves a partial vacuum, as no air can enter from below to fill the space of the displaced water. The consequence is that the softened particles of clay are drawn together, and by cohesion the clay becomes a solid mass. Of course, the interspaces thus obliterated diminish the volume of the clay, and this is manifested by the gaping cracks that show themselves in the clay, dried by surface evaporation. These cracks measure the amount of consolidation in the drying· clay. Now if the water can escape below, the air follows the descending water and fills the pores of the clay as the water vacates them, and consequently there is no consolidation of the clay. The admission of air from above is an important factor in the beneficent influence of tile drainage.

2nd. This free ventilation of the soil is the secret of the effect of tiling in time of drouth. A finely pulverized soil has a power of condensing moisture from the atmosphere that circulates freely through it. This can be readily demonstrated by experiment. Take about five pounds of good soil, well dried and finely pulverized—put it into a small basket after carefully weighing it and expose it in some unsheltered place, on a clear summer night. Weigh it again in the morning, and perhaps you will be surprised to find that the five pounds of soil has gained from one to two ounces in weight from the water absorbed. That is apparently a small gain, but when we apply the ratio to the area of an acre of fine soil a foot in depth, it will amount to barrels of water. The capacity of air to hold moisture is measured by its temperature. Of a clear night the earth loses heat by radiation very rapidly, and the air entering it, whether from above or below, is soon reduced to the temperature of soil, and of course deposits the water it was able to hold at a higher temperature, on the same principle that dew is deposited on grass when it is cooled by radiation of heat. At the depth of three feet ;or more this deposition of moisture is chiefly in the day time, as at that depth surface radiation of heat has but little effect. If the clay is rendered porous by a system of through drainage, the air will penetrate it from every direction, and whenever the earth's temperature falls below the temperature of the air, water will be deposited if the air is saturated.

There has been much speculation on this subject, but theories apart, every farmer knows that in a drought the hard road side will dry to a greater depth than the adjacent cornfield that he is cultivating. Last year's drought demonstrated that properly tiled grounds produce fair crops, while on undrained clay land the crop was nearly a total failure.

3rd. There is another advantage in drainage that is of prime importance, though it is generally overlooked. The fineness of the particles composing a soil is an important element of fertility. The fertility of bottom land, especially that which has been deposited from back-water, depends chiefly on its extreme fineness. Now, in a cultivated field, where the water which falls on it is carried away by surface drainage, either with or without open ditches, it is always muddy—indeed, loaded with the fine particles of soil held in suspension. If that water had been filtered into an under drain it would have been as clear as a mountain spring."

If we stand by the roadside, or by open ditch, in time of a freshet, we see the torrent of muddy water carrying to the streams and lowlands tons of the best part of the soil.

How long can lands endure this waste and not lose their vitality, becoming what is known as worn-out lands?

Other, though perhaps lesser, benefits follow drainage, and have not only made it desirable but profitable. Among the benefits above referred to, and the first in importance, is that of lightening up the soil; a condition which allows it to absorb more moisture in times of heavy rainfalls than the more compact soils of under-drained lands would do. To the extent this is done, drainage is desirable; but with this small benefit, a large and serious loss to soil and crop is entailed by the drains immediately carrying from off and out of the lands all the water of the fall and spring rains and the melting snows of winter, over and above the small percentage held in the soil. Not alone is the surplus water lost, but it, in its course down through and away from the land, carries with it elements of fertilization natural to the soil, or contained in the rains and melting snows, as well as a large, and the most valuable parts and elements of the manures and fertilizers applied to the lands by the tiller. These escaping waters will in nearly every instance be found to be a strong, powerful, and valuable liquid manure.

In districts where tile drainage has been extensively adopted, it has been found that springs and wells in many instances fail.

This is an inevitable result of direct drainage whether on surface or beneath it. It is the surplus water held in marshy places, or sinking to some subterranean pool which furnishes the supply and keeps them living. Mr. Cole by his method drains the surface of producing soil perfectly, and gathers the surplus water in storage to feed soil and crop by capillary flow, or spring and well by percolation.

CHAPTER IV

The New System.

The principles and processes of this new system are so simple and plain that the farmer of average intelligence can not only understand and apply it, but, having so conformed soils as to set the system in operation, he may leave it to run itself, which it will do, day and night, year in and year out, summer and winter alike; and so perfect is its work at all times, that it results in the utmost possibilities of production.

The system is alike applicable to flat or hill lands, and its operation on one is as certain and successful as on the other. It is applicable to all soils, adapted to lands in and on which are found stone, or soil free of them; to arid districts, or to sections where the rains and snows of fall, winter, and spring are ample. For each of these conditions the principle and results are the same; some minor points as to materials used in construction, depth of drainage and of soil over the stored waters, being the only changes to consider. These we will treat under proper heads. For a general description, we will give it as applied by Mr. Cole on his hill-side, where stone were abundant and used in construction.

A trench, one yard wide and four feet deep, was opened along the hill side, crossing the plot a distance of about twelve rods. Two rods below this and parallel with it, another of the same dimensions was constructed, and so on down the slope. At the bottom of these were loosely placed cobble and blocky stone, until the trench was nearly half filled, then smaller stone were laid over these, finer ones then added, until a comparatively even surface resulted. Next came shingling with flat stone, two layers from end to end, breaking joints, being required. Over these was placed a cover of sods, weeds, straw, or best available material, to prevent the fine earth from filling the crevices between the stone. The excavated soil was then spread over all, the clay being the first returned to the trenches. The remaining surface and subsoil was pulverized and admixed, all fine stone being raked out, when the trench was filled to a point slightly above the original surface. This was done so, when settled, all should be even.

As these trenches were dug they were connected by overflow or drainage trenches, three in number, one at each end, and one in the middle. These were sunk in the soil about eighteen inches, and being filled with fine stone raked from the soil to a depth of about six inches, shingled and sodded the same as the reservoir, till the entire hillside presented a uniform surface throughout. This conformation of soils results in two water-tables beneath, one of percolation from the bottom of reservoir trenches, the other of overflow and filtration in times of surfeit. The lands thus fitted leave shingling of reservoirs and overflow below

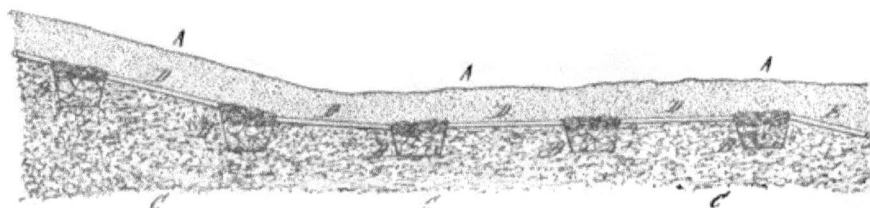

AA. Surface soil
B. Trenches
C. Subsoil
D. Overflow trenches
E. Outlet or drainage trench

Patented July 22nd 1884

No. 1.

the reach of spade, plow or other tools used in working the land, but left ample room for the roots of all growing vegetation. These trenches completed and connected, formed elongated reservoirs, which filled by the

water courses cut off, or by the melting snows and early rains, thus holding in storage thousands of barrels of water which would otherwise have
been lost in wash or drainage, and at the same time so saturating subsoils as to convert them into store-houses filled with moisture to be found,
at all times available for use in the growth of vegetation. If the reader
will take the above brief description and consider it carefully in connection with the cuts illustrating the system he will find no trouble in understanding the principles of construction.

The cut on page 34 will show how to arrange the system on
hill-side or sloping lands, when stone are plenty for use in trenches
and over-flow drains, and the soil is underlaid with hard-pan, or any
other firm subsoil.

No. 2.

In the absence of fine stone in the soil to be used for overflows,
gravel from the pit or other sources should be used and if neither are
available, tile should be substituted. We prefer tile.

Cut No. 2, illustrates the system when applied to very steep lands
which are terraced, used either for grapes, small fruits or ornamental
shrubs; it shows arrangement of narrow terraces with a single trench, or
wide terrace with double trenches.

Level lands are laid out, trenched and connected in the same way,
the only thing necessary being some point from which the surplus waters
can be drawn off at the desired depth beneath the surface soil. This outlet may be made from anyone of the trenches in the system. The waters
in the trenches cannot raise above the overflow or drain trenches, and
when below them, distribute themselves evenly through the subsoil. If

there is no point where drainage can be readily secured, sink a dry well to where the water will pass off.

Use of Tile.—The use of round tile in the system of sub-irrigation is suggested to meet the wants of those who may wish to provide for the sub-irrigation of small tracts where the soil is level and stone are not to be had in sufficient quantities or in any quantity worth mentioning. Also in lands not underlaid by retentive subsoils,

The greater ease of making round tile and preserving the form perfectly, leads us to favor (for the purpose of sub-irrigation) the use of large round tile, 6, 8, 10, 12, or 15 inches in size.

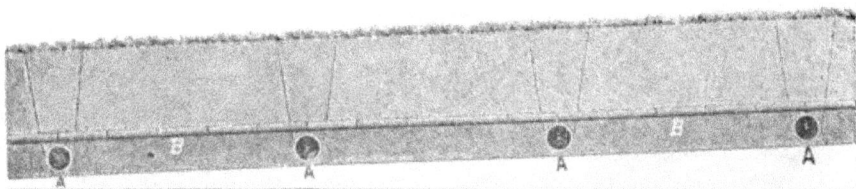

No. 3.

Cut No. 3 represents tile as used.

A A represents the large tile reservoirs and B B the drainage line of tile laid above and across the reservoir tile. The large reservoir pipe should be laid on a level and closed at each end. The water passing out slowly through the joints of the pipe. The clay near the joints will soon become puddled so that the water would percolate through it slowly when laid in clay. If laid in sandy, light or porous soils, the joints should be puddled sufficiently to prevent rapid escape of the waters. A small escape at the joints in addition to that which will pass through the porous tile, will be ample and effective. At the point where they cross it is intended that connections be made by means of openings in both the reservoir and overflow or drainage line of tile. The latter may be constructed of three-inch tile and laid at such distances apart as the character of the soil may require.

Square Trenches.—When small plots are to be fitted, or the nature of the soil is such that there is danger of the trenches filling with silt, a square tile can be used for the storage trenches. The trench can be opened and the top plate taken off, and the silt thrown out with comparative ease.

In sections where lumber is cheap, any planking which last well beneath the soil may be used. Round three-inch tile will here also answer for overflows.

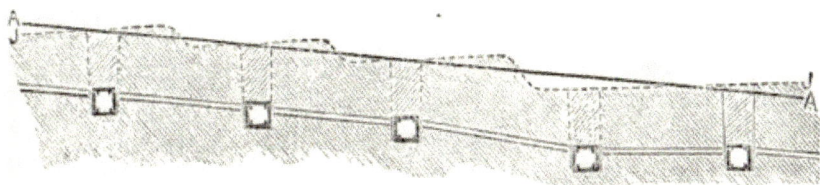

No. 4.

Cut No. 4 represents the arrangement of the square trenches on hill-side and flat lands.

Cement Trenches.—On all sand lands, when the sand is deep and leaches rapidly, trenches are opened, and the bottoms and sides made nearly tight by spreading a thin layer of mortar over the bottom, and up the sides as high as desired. The mortar is made by mixing one part Portland cement and seven parts sand. To cement the bottom, set a curbing of loose boards and pour in the mortar and level to the desired depth. Cement sides in same way.

The foregoing description and cut illustrate fully the methods of applying the system.

The next thing to consider is depth of soil to be left over and above trenches and overflow drains. Not only is it best to vary this according to the different uses to which the land is to be put, but as different soils have different capillary powers, depth must vary accordingly. For general purposes, we would recommend the following depths at which to put the overflows and drains.

In muck lands................................. 18 inches.
In garden soils .... 20 "
In sand   "   .... 12 to 15 "
In clay   "   .... 18 to 20 "

When lands are laid out for special purposes, the character of the growth to be made upon it must be considered. For all shallow-rooting vegetation the soil should have less depth than for deep-rooting.

For shallow growths the soil should be:

In muck.......... Not less than 15 inches nor over 24 inches.
In garden soil...... "   "   20   "   "   30   "
In sand soil...... "   "   10   "   "   16   "
In clay soil....... "   "   18   "   "   26   "

The following table is a good guide by which to go in laying out lands:

Muck will lift water in about 30 days.............. 18 inches.
Garden soil "   "   "   "   .... 28 "
Sand   "   "   "   "   .... 10 "
Clay   "   "   "   "   .... 28 "

Depth and Distance.—The depth and distance apart of the storage trenches are not governed by arbitrary rules, but the following general rules should be observed:

In sinking the main or storage trenches, the bottoms of them, regardless of material used, should be sunk below the freezing point, and should be from end to end at a water level, and when this is done the stored waters are kept several degrees warmer than the surface soil and the air above.

Evaporation from them keeps the soil from other than very shallow freezing; also helps it to warm some weeks earlier in the spring and permits it to be worked from two to four weeks sooner than it otherwise could. Size of storage trenches, or storage tile, and the distance apart at which they are placed, will depend upon the quantity of water desirable to store.

On lands where the water supply must depend wholly upon the stored waters of winter's snows and early spring rains, more trench capacity is necessary than when the system can be supplied by some regular, unfailing source.

Lands used for family or market gardens, where crops follow in succession the season through, lands devoted to small fruits, especially strawberries, fancy lawns, flower-beds and small shrubbery, will require more water than lands devoted to grass, grains, potatoes or other general farm crops, and should have more storage capacity.

The closer the storage trenches are placed together, and the greater their size, the more water they will store, and there is no danger of having this quantity too great. If, in applying, the storage trenches are put at greater distances apart than is desired or desirable, additional trenches can be at any time put between the original ones, and connected with the overflows. In putting in storage trenches, we would advise the setting at either end of the trench, at the head of the system, a wooden box, about six inches square, or a six-inch tile, rising about a foot above the surface of the ground.

Through this may be seen how much water is in the trenches or waters charged with certain elements, or for certain purposes may be run into the system. If a large territory is covered put them at other points.

For *Family* and *Market Gardens*, small fruit growing, etc., the system should be applied with a view to storage of all water possible, that they may be ample to furnish all the moisture crops of this kind require, and that every particle of manures and fertilizers put upon the ground may be converted from solid to liquid forms. Put the storage trenches in as close together as possible, or as you can at first afford to do. Carry the bottoms below the freezing point, and make overflow connections ample to carry all the water from trench to trench without rising into the soil above.

Put in, at the head of the system, the upright box or tile before mentioned, that you may run into storage all the waters from roofs of buildings within reach of the plot ; also liquid manures and house-slops, which should be carefully utilized. Free all liquids of their solids before letting them into the storage trenches. Establish as near the head of the system as possible a mulch vat, made by setting on edge boards to the desired height, staking them firmly on sides ; no bottom is necessary. Into this vat throw everything of a mineral or organic character useful for manure that can be procured. Stable manure, weeds, muck, leaves, night-soil, leather scraps, tobacco stems, lime, ashes, plaster, bones, and bone dust. There need be no fear of losing ammonia by adding lime. The lime is needed for the rapid decomposition of the manure.

With every rain leaching takes place directly from this vat into the storage trenches, and the water is rich in plant food.

All solids too heavy for capillary attraction to lift, are deposited in the soil above the trenches as the water passes down through it. For vegetable and fruit growing, the soil over the trenches should be from 18 to 24 inches in all soils other than sand. This depth gives deep rooting where necessary, and to shallow-rooting vegetation ample moisture by capillary flow. Soils filled with stone should be deep plowed, or better still on small plots, spaded and forked, that the stone may all be taken out, whether used for construction of other trenches or not. Stone in the soil, especially those near the surface, interfere with the proper rooting and downward growth of plants, and under some conditions, cause fungus and other plant diseases, being particularly fatal to strawberry plants. Removing the stone and working the soil tends to giving a compact, fine producing surface soil which holds and draws waters in capillary flow more abudantly than coarser soils, or those filled with stone, and plants of garden and small fruit varieties require an abundant supply of moisture.

Lawns.—By this system lawns can be carpeted with a fine, velvety green grass, starting earlier in spring, and never ceasing in growth until killed by the frosts of winter, and when well protected by snow will remain green all winter. Put in storage trenches as is done for garden purposes in every way. Turn into them all the waters which can be so led, and supply whenever desired liquid manures. This saves surface dressing at any time other than in winter, when the manures put on will protect the grass, and the leachings of it will be deposited in soil, or stored in reservoirs to feed the growth of early spring. Over the system here the soil should not be more than twelve inches, as grass thrives best on soils cool and at all times, full of moisture.

The more generous the water supply to grass the better the growth and color, as long as the supply is moving, and not stagnant water. Flowering plants and shrubs, when thus supplied with food and drink, reach perfection in growth and blossom.

Orchards.—Fruit trees make rapid growth of wood, shed their old bark, become freed from scale and the insects they harbor, have few if any off years, put out more blossoms, develop more fruit, and ripen it off in beauty and perfection. Disease is less likely to attack a tree, and if it does so, can be much more readily and perfectly checked. Hibernating pests which infest fruit trees, stinging blossom and fruit, will not enter and remain in the moist soils in any number, but should they do so can with their spawn, be readily destroyed by special applications carried through the soil by the circulating waters.

In orchard, and among fruit trees of all varieties, the trenches or tile should be as large and of as great storage capacity as possible. Trees should be as far apart as the greatest distances recommended by nurserymen, and the trenches located between them—a single trench to each row and that exactly in the center of the space between rows.

When too near, the roots of a tree are apt to put out a mass of small, sap roots, which enter and choke up the trenches. Give them distance, and the power of capillary attraction will draw all the necessary water to the tree, and the root-growth be checked. Of irrigated orchards, Mr. Henry Stewart in his valuable book says :

" It is doubtful if there is a single orchard or vineyard in the United States, except California, Utah or Colorado, subjected to a systematic irrigation. At the same time it is doubtful if there is any country in the world in which irrigation could be more profitably applied to fruit culture than here. The experience of orchardists proves that drought is accompanied by destructive attacks of insects. How far these depredations might be prevented by irrigation cannot be predicted, but it is beyond doubt that the vigor of growth that would result from a sufficient supply of moisture to the roots would greatly mitigate the effects of these attacks. The apple trees that never have an ' off-year ' are those grown near bodies of water. A California vineyardist who irrigated his vines immediately raised his product to eight tons of grapes per acre, and greatly improved in quality."

When the supply of water to the system depends wholly upon the rains and melting snows, there is no danger of the amount being too great for tree and fruit, as the annual rainfall is never in excess—even when all held—of the amount trees require. If other supply of water is provided, care should be taken in its use, as the penalty for an excessive irrigation is a crop of fruit of inferior quality, watery, soft, and without flavor. A good rule is to irrigate four times a year only—at the starting of the first growth, at the blossoming, at the setting of the fruit, and at the period when the fruit commences to color.

Grass, and other general farm crops, do not require the careful preparation of soil necessary and advisable for gardens and small fruit lands.

In applying the system for ordinary farm crops, give as great storage capacity as possible, as the more water stored the greater the benefits. When the storage trenches are completed and the connections for overflow made, return the excavated soil, plow the entire plot and plant or seed as desired.

As the system receives, holds, and returns to the crop the waters of rain and snow, a complete change in the character of the soil, its color, temperature and power of production will be seen to gradually and surely take place. This change is absolute and sure, caused by the soil's filtering out animalcule and other elements ; the fertility of the waters is, as it were, screened and strained into the land, enabling it to nourish vegetation.

Grass abundantly supplied with water is one of the most profitable crops grown, requiring much less labor and always finding a ready cash market. This system grows grass to its utmost perfection and limit of production, as it enables the waters to enter the soil and holds it there, keeping it cool and supplying the constant moisture grass requires for perfect growth.

Arable Lands.—As on these lands a water supply must be had, the necessity for water storage does not exist in as great a degree, and the tile or trenches for distribution of the water beneath the surface may be larger or smaller—according to judgment. Here size must be governed by extent of tract to be irrigated and amount of water to be carried beneath it.

Lay out the grounds in parallel trenches, and in these put the distributing tile on a water level, and connect at alternate ends. By an upright tile at head of system provide for running the water in, and at the foot of system close the tile that the water may be held and forced out by its own pressure at joints into the soil. Eighteen inches in all but very sandy and gravelly lands is a good depth at which to lay the tile, as this permits the soil to hold sufficient water for the crops, without sinking below the power of capillary attraction, or coming so near the surface as to interfere with tillage or be subjected to immediate evaporation.

In rainless sections, or districts nearly so, evaporation is much more rapid than in others, and the loss of waters supplied in surface irrigation is a very large percentage. Put your water below the roots of crops and evaporation is in very large part checked, and the land never bakes, but keeps cool, moist and loose. One-fourth the water used underground is better than the whole on the surface, and the saving which can be thus made will alone pay all costs of application. Where the rainfall of winter is copious, by constructing storage trench of large capacity the rains may all be held for spring and summer growth of crops.

Artesian Wells.—In Texas, Florida, Colorado, and many of the other Southern and Western States and Territories, artesian wells are readily obtained and may be used when necessary or desirable as feeders to this system. When they are used and wholly depended upon for water supply the system can be put in with smaller storage capacity, but in every other regards the same.

Waters from these wells, as from springs, etc., are never of the benefit to crops and soil as are the waters of rainfall, having been in their percolation and flow through the minute water veins of earth, deprived of the most valuable of their fertilizing elements, and having in many instances been charged with mineral and chemical substances, the washings of which are injurious to plant life. When artesian well, spring, or even river waters are used, it will be found necessary to use greater amounts of manures and fertilizers than where the rain-waters are fully stored, and changes in the nature of the soil will never be as great. At some points artesian wells are found whose waters are of a very high temperature, and here very advantageous use can be made of them, as their waters can be made to impart early warmth to soil and continue it until late in fall, in some localities the year through, in fact, and that to a point where plant growth may be made every month in the year. In another chapter, devoted to the use of hot water in this system, will be more fully explained the benefits which can be derived from the use of these warm waters from artesian sources.

Swamp Lands.—These lands, in nearly every instance, are so located that they do, and have for generations, received the waters and washings of higher lands. To the fertility thus brought to them has been added from year to year the decay of the rank vegetable growth which has grown upon them, and they are thus made the most valuable of lands when once reclaimed. On such lands as these can this system be established in one of its most valuable forms. Lay out the line of storage trenches, and in mid-summer, at the time of drouth if possible, sink them four feet or more. Fill with stone to within two feet of surface, or if stone are not to be had, lay a large tile. Connect storage trenches by the overflow drains two feet below the surface, and put in the drain which is to lead off the surplus at the same depth, tapping the trench which is nearest the point where the surplus water is to be led off. Replace the soil over all.

When the rains come, or the waters wash down upon it, they immediately sink into storage until all trenches contain two feet of water, having risen to the point of overflow. Above this point only water of capillary attraction can flow, and the two feet of soil above is in the best possible producing form.

If the swamp was formed by springs a water table is created for them, and they will keep the water supply up at all times.

CHAPTER V.

Workings of the System.

In the considerations of the workings of this system we will first take up that of irrigation.

The trenches and overflow drains having been completed and the soil replaced over them, the workings of the system begin by the trenches filling with the waters cut off in underground flow, or by the leaching down through the surface soil of the melting snows or falling rains. Having once filled to the point of overflow, all surplus water is carried off. As long as the rainfall is ample, or the soil is filled with water the trenches will remain full, or nearly so. When the rains cease, and the growing crops and evaporation begin abstracting the moisture from the soil, capillary attraction and the surface phenomena of evaporation begin the work of lifting the stored waters from the trenches, or the saturated subsoils beneath and between them.

During the growing season, evaporation is in excess of the rainfall with reference to bare soil, or with reference to transpiration from plant growth. Consequently, the direction of movement of the soil water in this climate is upward rather than downward; not to exceed ten per cent. of the waters entering the storage trenches, or the subsoil beneath or between them, will, in ordinary soils sink to a point where the powers of capillary attraction and evaporation cannot pump them back to the growing vegetation. By this system a water table is artificially created which will make farming a success, by giving a stored water supply within the soil from the surplus of the season when evaporation is somewhat checked, or from the waters accumulated from surface or saturated flow. This water table, or storage, is of great consequence to the tiller of the soil, as furnishing the source of prevention or mitigation of drouths.

This water table must be uniform, and controlled through the process of draining, or we would have conditions unfavorable to the working of the land and to plant growth, for vegetation cannot thrive with roots in stagnant water.

The series of overflow and outlet drains used in connection with these storage trenches furnish this drainage, and establish a line above which water of upward flow cannot pass, and below which sinking waters must go. We hence have a depth of soil above the trench and drain level which is freed from the washings of the waters of the storage waters, and which is kept from the stagnation produced by the action of water in excess; it hence retains and increases its power of capillary action, and draws regularly through this capillarity from the stored waters, never abstracting in excess of the plant needs, and in proper soil and under proper conditions of soil treatment, never stinting the water supply to the plant. This ever upward movement of water, only intermittingly interrupted through percolating water of rainfalls, conserves soil fertility within itself, and even recovers soil fertility from the layers of soil subjected to the steepings of the reservoir water.

These facts suggest a direction as to the depth at which drains should be placed. A drain should not be so deeply located as to carry the water table beyond the power of capillarity to freely convey to the surface the stored waters, but should be as deeply laid as consistent with this purpose of securing water movement to the surface.

The capillary flow of the stored waters can be largely regulated by two simple methods. Where a crop is put upon the land to which a large supply of moisture is necessary, compress the soil closely beneath the seed or plants. Compression of the soil re-establishes capillary connections which become broken through the process of plowing and fitting the land, and hence is of avail toward supplying a constant access of moisture to the young plant. The firmer the soil is made the more rapid it draws the waters. Should less moisture be desired than the soil naturally draws, straw or coarse manure may be worked into it to an extent sufficient to check the capillary flow to any point desired. These conclusions can be readily verified. During the dry season of July let coarse manure be spaded into the soil in preparation of a strawberry bed, and then immediately set out the young plants. Unless timely rain intervenes, it will be found that these plants will perish from drouth, as the coarse manure has broken the capillary connections, and no supply of moisture can come from below.

If however, this land be thoroughly rolled with a heavy roller, by which contacts are established and connection with the lower soil furthered, capillary attraction will furnish a supply of water from below, and the plants will receive sufficient nourishment for their growth.

CHAPTER VI.

Effects of the System.

On Soil. The effects upon soils of this system are most marked, especially so upon clay soils, soils underlaid by hardpan or sandy soil. Clay soils are above the overflow, and drains trenches deprived of their naturally sticky, cold, and wet condition in spring and fall, are early made light, dry, warm, and friable, and put in the best possible condition to work.

In heat of summer evaporation from below keeps the soil moist, prevents baking and cracking, and the ground is at all times soft and easy or tillage. The percolation of the waters from level to level, and from trench to trench, carrying with them in their flow, air, light, warmth, and elements of fertilization natural to the waters, or taken up and deposited by them in their passage, so change the soil between the trenches that it gradually becomes as the surface soil, sweet, warm, light, and rich in plant food down to a point as deep as the bottom of the storage trenches. This condition of the soil allows the roots of growing crops to more freely penetrate it. There they ramify, and in times of drouth are cool and moist, independent of the scorching sun and drying air above. The foliage remains green and thrifty, perfect growth, development or fruitage goes on, being constantly fed by the life-giving moisture.

Again, the soil being thus opened the aerial gases, especially oxygen, enter it, and decompose the organic matters so that they can be taken up by, and nourish, the plants. By opening the soil we admit the air.

Air is as vital a necessity to vegetation as water, and if access of air is denied the roots of the plants must perish. Where water goes air follows, and as evaporation takes place air fills the space previously occupied by the water. It equalizes the temperature and prevents sudden changes of heat and cold. It renders the soil drier and warmer, earlier in spring and later in fall, thus greatly prolonging the planting and growing season. In dry seasons the soil is moistened by condensation of the air admitted into it not only from above but also from the drains and trenches beneath, as well as by the stored waters. The soil is thus made more open and mellow, roots penetrate farther and get more nutriment. The open dry soil absorbs miasmatic gases which enrich the soil and purify the air, thus increasing health and wealth.

These same changes and effects are produced on hardpan lands, and firm retentive subsoil. Sand lands are made moist and cool, and given the power to nourish and feed vegetation through the medium of the waters.

Perfect vegetable growth on sand in heat of summer is not possible, as the temperature of the sand becomes such as to literally burn it root and branch. Surface irrigation or watering affords but little relief, as scalding then but takes the place of burning.

Apply the waters beneath and all is changed.

They are beyond the direct effects of the sun and its rays, and afford coolness and moisture at the same time. Sandy soils warm much earlier in spring, hold heat later in the fall, and where they can be kept cool and moist in midsummer can be made to bear continuous crops for longer periods, and in the most luxurious growth possible to vegetation.

Poverty of the soil (a condition, indeed, mainly due to a poverty of the water) is unknown, and impossible under the system. Any element lacking in the soil to the growth of a crop can at once be carried into, and through it by the medium of the stored and circulating waters.

Lands now said to be worn out, and which, through their firm and compact condition are not capable of absorbing air, light, or water in quantities sufficient to support vegetable life, can, by this system be made to produce crops in as great abundance as at any time in the past, and this in many instances, without the addition of manures or fertilizers to them. Open them up by it ; allow them to drink in fully the season's rains, absorb the heat, take in the light, and beathe in the air, that all may reach the roots of the crops to be put upon it, and it will in most instances be found these elements were alone lacking.

Washing of the soil by heavy rainfalls with their loss and damage to lands and crops, is absolutely prevented.

The moist, open, and light soil above drain and storage trench at once taking up the rainfall without regard to its extent, and pass it down to the reservoirs. The surplus if any will be passed off through the overflow trenches without in the least disturbing the surface or the crops grown thereon.

Soils under this system are also made untenantable for destructive insects and other pests which infest or hibernate in them, and which every year in one portion or another of the country, reduce farmers' profits or cause them to disappear. Moisture is destructive of the pest and its spawn, but when this system is in use this condition need not be depended upon alone to rid the soil of vermin. The tiller has it in his power to change the circulating waters of the system with elements and remedies which will at once destroy or drive out every pest it may contain. To illustrate : If a plot of ground becomes infested by wire, cut worms, or other white grub, they can at once be destroyed by letting into the storage trenches a solution of copperas water. This is taken up by the stored waters and carried by percolation and capillary attraction all through the soil, the pest destroyed, and the crop benefited, as copperas is a powerful stimulant and fertilizer. Kanite, saltpeter, muriate of potash, wood ashes, and many other substances destructive to vermin but highly beneficial to vegetation can be used in the same way.

Hibernating insects will not enter or remain in soils charged with some of these salts, as they do in soils which are not. Where the land has been charged, potato bugs, circulio, and insects of their kind, have largely disappeared.

As these things are used to destroy vermin, so may others be used to cure or prevent blights and other diseases which attack plant, crop, shrub, or tree.

By the aid of the stored waters can the necessary food or remedy be carried directly into the circulation of the object to be treated. Many of the blights and diseases to which fruit trees and shrubs are subject, and which no outward application reaches, can be at once checked or cured if the remedy is carried directly into their bodies, limbs, and foliage, by medium of root and sap. As internal remedies are necessary to the system of man or beast in case of disease, so are they necessary to plant life, and having the power to administer, you have the power to cure or prevent.

When special applications are to be made for a special purpose, dissolve the ingredients in water, and lead it into the storage trench, directly above the soil, crop, shrubs, or trees to be treated.

Effects on Crops. The effects of the system on crops are as marked and wonderful as they are certain. As it warms and dries the lands some weeks earlier in spring it permits of much earlier seeding or setting, and gives the crop a good early start, which every cultivator of the soil knows is the making of it. So soon as the seed sprouts, or the set plant puts out its first roots, the crop begins to receive and take up all the water necessary to its perfect growth, development or fruitage. As long as the spring rains are abundant for the demands of the crop, the stored waters are not drawn upon; but let them fail, and the demands of the vegetation exceed the supply furnished by the rains or contained in the upper soil, and capillary attraction sets her countless pumps at work, drawing to the roots of the crop from the stored waters beneath all the nutriment necessary to its continuous growth and perfect development, but never in excess of the demand.

As these stored waters rise in capillary flow they bring with them in solution all the fertilizing elements natural to the rain and snow waters, also that which they have gathered from the soil in their percolation down to storage. Plants and crops thus watered and fed become more perfect than is possible under other conditions, grow to a much greater size, and when of the flowering and fruiting varieties, have the strength and power to put out a greater number and stronger fruit-stems and blossoms, and of carrying them to perfect growth. Plants, shrubs, vines, bushes, grass, or any crop which stands upon the ground through the winter, receive constant and never-ceasing benefit.

As evaporation from the stored waters prevents to a large degree freezing of the ground, and keeps the soil some degrees warmer a few inches beneath the surface, than is the air above, the crop, plant or shrub, put out and build up a mass of white root which will begin to feed and water them as soon as warm weather starts the sap. This growth of white roots, in their formation, exert the power of capillary attraction to a greater or less degree, and draw to them the waters in downward or upward flow. These waters bring and deposit food and fertility, which the plant will at once take up and assimilate, as soon as growth above ground begins. In this feeding upon substances brought to them, they make their own selections, choosing that which is necessary or best suited to their growth and development, and rejecting everything offensive or unsuited to their natures.

On the plot of ground put under the system by Mr. Cole, this winter growth of root has taken place every year, and the rapid, strong growth shown by the vegetation occupying the ground in early spring is proof that great strength is accumulated by and for the plant during non-producing months.

CHAPTER VII.

Builds Up Soils, Saves Manures.

Artificial watering by any method never gives as satisfactory results, even when ample, as will one good shower of rain. This is because the waters of spring, well, or stream, have been strained and deprived in their flow through soil or along water course of the elements which they, as rain water, originally contained, elements which renew and build up soils worn or wearing by constant cropping. Rain water becomes so chemically changed and charged by the process of evaporation, condensation, and fall, that when it comes in abundance crops grow, thrive, and develop to a state of perfection not possible even with any other equally abundant water supply, and the soil shows evidence of having been enriched.

To the chemical changes which take place in the rain waters in their rise by evaporation and fall as rain are the additions of solids which they gather and deposit when confined.

These solids are concentrated fertilizers, far superior for plant growth to anything man has ever been able to compound. They are also much more abundant than is generally known or supposed, a year's rainfall making a deposit of no mean amount over a surface where caught and held.

As it is with the rain waters, so is it with the snows in their effects, elements and deposits.

While all snows contain these ingredients, and are of value, the snows of March are particularly rich in them, and are of sanitary and medical value to man, as well as soil and crop. While discussing with Dr. Thos. Herron, a well-known physician and specialist, of Cincinnati, Ohio, this question of storing the waters of rain and snow, he remarked: "Yes, in thus saving for your crops these waters, you gain an advantage which no other method can give you.

Not alone for soil and crop have these waters wonderful chemical properties, but for the use of mankind as well. The snows of the month of March contain elements, electric, magnetic or chemical, we know not which, which are of particular value.   One of the most popular and successful eye waters on the market is nothing more or less than water of March snows carefully gathered (so as to save filtering), melted and bottled.

Again, if the ladies would so gather and bottle this snow water, using it for the purpose of sponging the face, they would find it, as a preserver of complexion and skin, much superior to the best known cosmetic in use.   That it has equally as many and valuable properties for crop use I cannot doubt."

Where the country is covered with forest, and the soil is made a perfect sponge by the roots, decaying woods, leaves, and mosses, the snow and rain waters are absorbed, and their deposits held in the ground. These and the small amount of vegetable decay annually taking place on them, are all the fertilizers the soil ever receives, and yet the yearly growth of wood, foliage, wild plants, vines and grasses exceed in weight and bulk any crop which could be grown on the same soil when under tillage. This primeval growth takes place year after year, and increases rather than decreases.

Clear off the forest and crop the ground a few years and all is changed.   Tillage breaks up the little water cells of the loose soil, packs it, forms by compact root of crop, sun-baked surface in summer, or frozen one in winter, a roof, so to speak, which turns off in large part these waters, leading them with their valuable contents to ditch, drain, or stream, where they are lost to soil and crop.   This loss is not, nor can be ever made up, by the application of barnyard manures or commercial fertilizers.   This is also not the only loss occurring under these conditions.   The flow of water carries in its course fine particles of soil, or manures in bulk or solution.

As fertilizing applications are generally made in fall, winter, or early spring, when the rains are copious, they, or the melting snows, carry into the streams a very large percentage of the most valuable parts of the applications made.   What farmer has not seen these dark-dyed waters rushing off over the surface of his land, carrying away the best of the soil, and his fertilizers also.   So great is this loss on the average, the checking of it would save enough in manures and increase of crop to pay the expense of doing all work required by the new system.   Tile drainage, which has in the past few years become popular, and is being rapidly extended, has in some degree changed these conditions and checked this loss by making the soil light and porous, enabling it to take up the rainfall. Where this is light, and the amount not sufficient to sink down to the tile, all is saved ; but where it is in excess of what the soil can absorb and hold, the tile at once leads the surplus away, and this too is lost, with what it still holds of plant food.

Here once more are the rains of fall, winter, and spring largely lost, as they afford a much larger amount of water than the soil can of itself retain, and the surplus escapes.

Rainfalls on tile-drained lands, which are of sufficient volume to sink down to the drains, deposit in percolation all, or nearly all, their solids ; but during this downward course they dissolve and carry with them near-ly the whole of every fertilizer or manure contained in the soil which can be rendered solvent, and would carry off all, should they be contin-ued for any great length of time. Go to the outlet of the tile drain beneath a highly fertilized plot of ground, and catch some of the escaping waters in fall or spring, or during a heavy rainfall in summer, and you will find they are rich in elements of fertilization. In other words, you will find it is a liquid manure, so rich in plant-food one would gladly pay its full value in hard cash to have it returned to soil and crop. Mr. Cole. by his system, prevents and saves this loss.

His lands are made light, open, and sponge-like, and have the power to absorb the rainfall in its entirety, summer, fall, winter, and spring alike, and having taken it all up, they pass it down to drain tile, and from tile to storage.

The importance of passing the rainfall down through the soil, rather than over the surface, is seldom thought of or in anywise appreciated. Stick a pin right here—when we say the water of rainfall that falls on the sur-face of agricultural lands ought to pass through the soil and subsoil for the better fertility and better mechanical condition of soil and subsoil.

Think of the fertility wasted annually by the surface washing of the land, the soil loaded with richness, and the finer particles carried away in the muddy water never to return in any form.

In percolation down to storage the water does not deposit all of its mineral and gaseous elements, which possess fertilizing qualities, but gathers to it additional richness, which it carries into storage, there to hold and steep in preparation for return to crop as liquid manure, a form in which it has the quickest, most powerful, and complete effect possible on vegetation.

The use of liquid manure is an economy, and results in a saving of time and labor, and increases the effectiveness of the solid manure. Being applied in the condition in which it will enter at once into the circulation of the plant, there is no loss of fertilizing matter. The crop, fed in its early stages of growth, receives its nutriment in such quantities and at such periods as will exactly meet its needs and force it into most luxuriant growth.

As these leachings of the soil and manures reach the storage trenches they commingle with the other waters contained therein and become diluted, and according to Mr. Stewart in his "Irrigation for Farm, Garden, and Orchard," should they become so much diluted as to run off perfectly clear, might be of sufficient strength for all purposes."

Discoursing on this subject of liquid manuring, Mr. Stewart proceeds further to say: "The danger lies in using liquid manure of too great a strength, rather than in diluting it too copiously. It is this fact which makes the New Agriculture the way, *and the only way*, to perfection in culture."

It has been found in practice when a heavy rain has filled the storage trenches in places where there was but a very small supply of manure, and the waters were not less than a hundred times weaker than ordinary liquid manure, they still give wonderful strength and color to the crops. A plant may starve on the most abundantly manured or fertilized soil when it is dry. In fact, fresh manures and all commercial fertilizers applied to crop or soil during, or just previous to a drowth, are injurious to the crop, as they of themselves furnish elements which burn out and destroy the vegetation. They may be applied at a time when all conditions are in their favor, and work benefit in the start, but dry out the surface soil and they change as to fire and destroy. Commercial fertilizers, when honestly made, are of great value to the cultivator of the soil, offering, as they do, in concentrated form and easy of application, some of the most powerful elements of fertilization known or possible.

That many who have used them have met with failure in realizing expected benefits is well known. This failure is owing not to lack of merit, or the necessary elements in the fertilizer used, but to method and conditions of application.

As an illustration : A man who is cultivating the soil under old methods and tile drainage, purchases a quantity of some standard brand of commercial fertilizer for use on soil and crop. He is given rules of application, among which is one made prominent and impressive. "In applying this fertilizer, do not let it come in contact with seed or root of plant, as it is so powerful in some of its elements it may cause injury to young and tender roots and plants." Such being the case, he must either apply before the seed or crop is put upon the ground, or between rows so far away that the direct leachings from it will not reach the root. Having so applied, if there were to occur just enough rainfall to slowly and continuously act upon it until all is dissolved and diffused through the soil, the effect would be all claimed or expected, but this never happens. If applied as above mentioned, before the crop, heavy rains wash the soluble parts in large percentage away either in surface wash, or by escape through the tile (where used) ere the crop can begin to receive benefits from it. If later in the season it is put between rows of crops, it is more than liable to lack the necessary rainfall to dissolve it in time to feed the crop and materially assist growth ; it is also liable to those sudden and very heavy summer rains which cause great damage and loss as do the more protracted ones of other seasons. To say that 50 per cent. of fertilizers and manures put upon the ground in ordinary conditions is lost is a very conservative estimate, and this loss is a very serious and expensive one.

Apply the same brand of fertilizers on a plot of ground under Mr. Cole's system and see results. If applied early in the season, the rains wash it down into the soil, and what the soil does not hold either in liquid or solid form, the storage trenches receive, dissolved and diffused to return to plant and soil by capillary flow when taking place.

If applied after the crop is on the ground, and at any time during the season, it has a soil made sufficiently moist by the stored waters (if not by rainfall) to surely and slowly dissolve it. If before this is fully done, a heavy rain comes, down into the storage trenches all goes, only to be returned in good time thoroughly dissolved. In this form its effect on crops is immediate and complete. As it is with commercial fertilizers, so it is with every manure used.

Mr. Cole found, after having had his system in use for two or three years, that the fertilization contained in the rain waters, and that which they gathered and saved from the soil, was, with very small additions made in the spring, ample to carry his crops through the year, and develop them to perfection. When he first made known this fact, it was said by many persons, and one or two agricultural publications, that his statement was not true, and that no method could be found which would lessen and save this expense of fertilizing. Since then an experiment has been made at the State Agricultural Experimental Station, which proves his statement true, that the benefits of the water, and the saving of manures will, in a very short time, pay all the expense of applying the system, even in its most perfect and ample form. We give an extract from the report of the experiment as made from the station.

"Effect of Commercial Fertilizers.

Upon the dryest and most gravelly portion of the field which the fodder corn was grown, several one-tenth acre plots were measured off and treated with several forms of concentrated fertilizers. The corn, Sibley's Pride of the North, was planted on May 7th, the fertilizers were applied broadcast on June 4th, 400 pounds per acre being applied in every instance. Below are the given results.

Plot.	Kind of Fertilizer.	Amt. per Acre.	Yield. Lbs. per Acre.
1.	Ground Bone..................	400	17,100
2.	Cotton Seed Meal.............	400	15,450
3.	Cotton Seed Ashes	400	13,900
4.	Equal parts of Cotton Seed Ashes and Cotton Seed Meal........	400	13,600
5.	Equal parts Ground Bone and Cotton Seed Meal...............	400	13,200
6.	Equal parts Ground Bone and Cotton Seed Ashes.............	400	14,730
	Unfertilized plot on similar soil, but in a moister situation.....		20,610

The corn was cut September 12th and was well matured.

As will be seen by the table, no results in the crop can be traced to the use of the fertilizer. This is undoubtedly due to the fact that there was not enough water at hand to enable the plant to use the fertility that was in the soil before the fertilizers were added. There are many things that go to show that lack of success with concentrated fertilizers may be entirely due to the peculiarity of the climate of the particular season in which the fertilizers are used."

CHAPTER VIII.

Use of Hot Water.

We will now treat of that feature of Mr. Cole's system when a steady stream of cold water, drawn from any source, is passed through a coiled pipe or boiler, and then emerging, is dropped into the trench, the stones or tiling heated, and when by surface protection the winter months are, to a great extent, made those of production. Let the following serve as an illustration:

"A curious experiment has lately been made at Acqui, Italy, by the proprietor of baths. The gentleman has at his disposal an inexhaustible supply of hot water from a natural spring, the temperature being 167 degrees Fahrenheit. The surplus not required for the baths has been diverted so as to flow through pipes to a garden on the outskirts of the town. Here the heated water flows beneath a number of forcing frames containing melons, tomatoes, asparagus, and other garden produce. The result is that a supply of these delicacies is ready for market at a very early period of the year, when, therefore, they fetch high prices."

Locate your coil or boiler at a point most convenient to your water supply, whether it be obtained from hydrant, steam pump, or from the lower trench of Mr. Cole's system established on a hillside. When the more tender and delicate class of vegetables are to be grown in the winter months, make the storage trenches from four to eight feet wide, and any length desired, and not more than four feet apart. Use stone where available for their construction, and connect by the overflows; cover with soil, which should here be from two to three feet deep, in order that there may be room for root growth above a point when the hot waters or steam will scald. Directly over the trenches build cold frames just as ordinarily constructed. Turn on your water, start your fire, and you will find growth can be made over them all through the year.

Immediately below your cold frames extend the system as for garden use, and let the hot waters flow through it.

On this ground very early growth of vegetables can be made, and a late growth continued in the fall. Here frosts are not possible, and only very hard freezing can do damage. The soil obtains heat as well as moisture from the steam and heated stone. We have said water could be taken from the lower trench of a hillside system.

The flow from such a trench will be found to be ample all through the months of autumn, winter, and spring, when the rains and snows afford steady supply. Mr. A. I. Root, of Medina, Ohio, publisher of the monthly journal, *Gleanings in Bee Culture*, has adopted and made use of the system, making a specialty of the hot-water feature. On the plot where he uses the hot water, it is supplied with the escape steam from a manufacturing establishment, and the crops are grown in the open air. In his journal, under date Dec. 18th, 1886, he published the following article under head of

"*Five Crops a Year.*"

"We commenced planting seeds and putting out plants, as you may remember, some time in February. Peas were planted in the open ground over these reservoirs February 1, and they yielded a crop of nice pods the fore part of April. Cabbage plants taken from the greenhouse and put beside the peas at the same date were, for a time, protected by a sash to harden them off, but after two or three weeks they too were allowed to stand without protection. They were repeatedly buried up in the snow, but the ground did not freeze, owing to the heat underneath. The plants made a vigorous growth, and were just as good as cold-frame plants, so far as I could discover. A few were left on this ground, producing heads of cabbage in April that weighed, some of them, ten pounds, and, as we received for them three cents per pound, we got thirty cents for a few single heads of cabbage.

After the cabbage we put on beets and lettuce, of which I have written you. The lettuce sold readily in May at five cents per head each. You will remember we started another crop between the rows when we were maketing one crop. After the lettuce we put out cucumbers which were started in the greenhouse. It took some little time for the roots to get down into the reservoirs beneath; but when they did they gave an enormous crop of cucumbers, which grew so rapidly sometimes that we picked them every day. These cucumbers, coming much in advance of any raised in the open air, sold readily at a good price.

After the cucumbers, we raised a fine lot of turnip-plants; and when these were put out in the open field the bed was cropped with lettuce, which now covers the ground, and the plants are so large and thrifty that they are an astonishment to every passer by. The year is not up yet, and will not be until next February; yet this ground has given us *five paying crops*, and many of the crops not only paid, but paid handsomely. Of course the ground was heavily manured. The two feet of soil that covered the stone reservoirs contained, perhaps, six inches of manure; but, my friends, will it not pay, and especially will it not pay for those who have *time* on their hands, with nothing to do? And this book, as you may remember, is written mainly for this class.

A squash vine, started in the greenhouse, and put out in June, made very slow growth for a time, but it finally started an immense squash which in time got so heavy that it broke the vine that bore it—said vine being run over a trellis. I told Father Cole about it, and he said I should have made a platform or shelf under the squash. Another one grew to an enormous size, and we finally sold about half a dozen in all from one vine. The quality was excellent.

Some time in the latter part of July or fore part of August, a pumpkin vine came up of itself. Nobody thought it could possibly ripen any pumpkins; but the single vine produced six, and ripened them all. The growth was astonishing. I presume vines of all kinds send their roots down at once into the reservoirs, and when they get there they make a growth that is really surprising. Some strawberry plants were started in March, and they bore a crop of fine, large berries, but we expect greater results from these next season. Pie-plant promises wonderfully. We have not tested the matter on asparagus, but propose to another season.

Such a plot of ground may be worked on not only evenings, but on rainy days. If it rains hard, fix a temporary covering of the shutters or sash, used in the winter time. Of course, the aid of exhaust steam, such as we use, will not very likely be at the command of many of our readers; but this exhaust steam is of no particular use from the first of April until the first of October; and by the aid of sash we can make plants grow nicely through March and October—yes, and through a great part of November, without bottom heat, especially where we have ground that is enriched to two feet in depth.

Perhaps the greatest satisfaction I have received in the use of the New Agriculture has been in making the swamp-holes and waste places on our eighteen acres the most productive spots we have. When I bought the place, a small stream, which we call Champion Brook, meandered across our low ground, traveling over more than twice the distance it would make in going straight across, as such brooks usually do. In fact, our whole creek bottom has been at different times washed over and cut up by this Champion Brook. Well, my first work, six or seven years ago, was to make Champion Brook go straight from where it came on to my land to the point where it passed off. This left the old creek bed, and in many places it had washed great holes down to the bed rock. These holes have been a bother and a nuisance until the present season. During the past summer, as I have told you, we have scooped out the contents, clear down to the rock in many places, and I tell you it was fun to do it—especially as the dry season greatly favored the work. These ho'es were then the receptacles for all the stone, tinware, boots and shoes, and everything that could be collected on the place, and during the past season we have been grubbing out the stumps, and using them to fill these gullies and holes. We make the surface level, then cover it with tinware, or something that will not decay, then down come the banks to the sides, until we can run both plow and cultivator right over them without any danger of striking or tearing up the stumps or tinware. Of course these holes are full of water, or, at least, as full as the outlets described in Chapter V. allow them to get. We want the water to come within a foot or eighteen inches of the surface of the soil, but no higher. Our crops over these places have been wonderful. I was pleased to see even friend Terry express his admiration and satisfaction at the looks of my potatoes, celery, and other crops. Celery seems to revel in such a situation. Our plants on the creek bottom are to-day looking as bright and green as they have been during any part of the year, and it is now the second day of November. Last week we dug some roots that weighed fully three pounds each. At five cents per pound this is fifteen cents per root, which makes quite a satisfactory return for the labor and capital employed.

In order to see whether our celery compared favorably with the best raised anywhere else, I have had samples of the finest from Michigan, from Cleveland, O., and even from the Arlington market gardens, near Boston. Some of the Arlington celery weighed three pounds to the root, but it was on account of the enormous suckers growing around the sides. I should call ours ahead anything that we have seen. Of course, fine tilth, with plenty of moisture, and not too much, are not the only things wanted; we must manure also; and I think it will pay well to manure the very best ground that can be found anywhere—at least such is my experience."

Here is evidence of the complete success and value of this system, and that there is profit in its adoption.

As the hot waters are used, so can, and should be, the high temperature waters of artesian wells, and where hot springs are found tropical productions should result.

CHAPTER IX.

Cost and Profit of the System.

So largely will the cost of introducing this system be governed by local and other considerations, that to name any fixed sum per acre is not possible or wise. We will only say it can and will be covered by an amount ranging from $30 to $300 per acre. When it is adopted in a section where stone are plenty on the ground, and it is put in for general farm purposes, and with abundant storage capacity, the work done in large part by the farmer and his boys, or regularly employed help, using his team and plow in starting, and, as far as possible, sinking trenches, the cost per acre will not amount in money outlay to the first-named sum, while the last-named will cover the outlay necessary to be made by a market gardener or grower of small fruits, who applies the system in its most perfect form, abundant in water supply, and in every way calculated to force production to the highest possible point in quantity, size, and quality. It will also cover cost of material to be used in construction where purchased, sinking of trenches, and preparation of surface soil.

In calculating cost of adoption, lay out your land in form as desired. Ascertain the number of trenches, fix upon the length, breadth, and depth, and figure cost of excavation. If stone are used in construction, and can be obtained on or near the plot, no expense will be necessary in filling, other than labor of gathering and laying in the stone in sucn way as to give the greatest amount of water space. If stone, tile, planking, cement, or other material must be purchased, ascertain the amount required and add to cost of trenching.

Add to these the cost of laying the material to be used, labor of re-covering the trenches and overflow drains, and you have the cost, approximately at least.

Mr. Cole, on his model, has made an outlay of a sum not exceeding $300 per acre, this amount covering in full the purchase and setting of the various small fruits, vines, and fruit trees now on the ground.

In the construction of the system, and fitting of the soil, every part of the work was done by hired labor, and a large part of it when wages were high, made so in that locality by the demand for labor which grew out of the opening up of a large oil and natural gas territory.

The annual sales of small fruits and vegetables now made from this plot of ground amount to a large per cent. in interest on an amount five times greater than the outlay made. Five acres are now under the system, produce in money value more than any fifty acres of land in Allegany County, and perhaps in the State. Space and time will not allow us to give an extended account of what Mr. Cole has accomplished, but here are a few general statements :

At the time Mr. Cole began his work, his home market was supplied wholly from outside sources with early vegetables and small fruits. He now furnishes the home supply in large part, and it may be said, "he is carrying coals to Newcastle," as he is now sending small fruits into the markets of fruit-growing counties of Western New York, the size, quality, and fine condition of his products giving them precedence over all others, and the demand for them has become so great he is not able to meet it, though pushing extension as fast as possible.

Competition under old methods of tillage has at various times been undertaken against him only to be abandoned after a short trial.

The summer of 1888 saw the small fruit supply of Western New York, with the single exception of Mr. Cole's, largely wiped out, caused by the deep freezing and soil heaving of the winter before, late frosts in spring, followed by severe and protracted drouths.

Contending against these combined adverse influences, Mr. Cole's system carried him through, and gave him a yield of crops much in excess of the best possible under old methods with all conditions in their favor.

Since adoption of the system, Mr. Cole has grown, one season with another, from 300 to 500 bushels strawberries per acre, many berries 8½ inches in circumference.

Blackberries per acre, 800 bushels.

Raspberries, per acre, from 400 to 600 bushels.

Potatoes, 1,200 bushels per acre.

Cabbage weighing 15 to 20 lbs. per head.

Heads of cauliflower, of the Snow-ball variety, in twelve days from the size of a walnut, to a perfect head 18 *inches in diameter.*

McLean's Advanced Peas, with pods 6 *and* 7 *inches long,* containing peas as large as *frost grapes.*

Ten dollars' worth of quinces, in one season, on a single bush, five years old.   Some of these quinces were as large as a *pint bowl.*

Three cuttings of timothy and clover in a single season, at rate of from three to three and a half tons to the crop.

The heads of the first cutting averaged 9 *inches in length;* the second, 8 *inches ;* the third, 6 *inches.*

Currants *doubled in size* and *yield.*

Evergreen sweet corn, the stalks of which measured 16 *feet in length.*

Apple and plum trees have stood under loads of fruit so great it was ound necessary, early in the season, to prop every limb and branch which could be reached.   All these productions have been perfect in every way, and absolutely free from blemish or disease.   Of these productions fuller accounts will be found later on, in reports made, and letters written by visitors who have examined into his work and its results.

We think our readers will accordingly admit the adoption of this system will be found eminently profitable to market gardeners and fruit growers ; in fact. will become a necessity, as in the active competition now existing, and which must always exist in the future, those who grow their crops under methods of the new agriculture will gain a substantial advantage over their competitors, and place themselves in position of security and independence of the vicissitudes of seasons.

In the home garden it will pay, since, on a very small piece of ground yearly, growths of fruits and vegetables can be made which will exceed the requirements of a large family.

When put in for garden and lawn, or applied to lands surrounding the house, it takes up all the stagnant waters, converts their solids to solution and gases, purifies them by causing their constant circulation, either by percolation downward through the soil and from trench to trench, or by capillary flow.   Moving waters are always living waters, and are never breeders of malerial and typhoid infections, as are those which are stagnant.

For general farm purposes its adoption will be found as profitable, as a rule, as for gardening. As it increases, improves, and assures crops on a small plot, so will it on large ones.

To no crops will it impart greater benefits and larger growth than to those of grass, potatoes, and apples. Every farmer knows that grass is the most desirable, but at the same time the most uncertain crop he can raise. It costs less to produce it than any other crop when the adverse influences can be obviated.

The two most fatal of these influences are drouth and heat, the one depriving the crop of the moisture necessary to it, the other searing it as by the breath of a furnace. Fortunately the farmer has it in his power. to overcome these by that system of subirrigation, which saves and stores water for use when needed, gathering and imparting nutrition to the roots of plants during the entire year, inspiring the utmost spring, summer, and autumn growth possible. On this subject we will once more quote from Mr. Henry Stewart :

" The permanent meadow is a very unusual adjunct to an American farm. Our climate is not naturally well adapted to the continued growth of grass. Our hot, dry summers are unfavorable. Generally it may be stated as beyond question, that the yield of grass is proportionate to the supply of water. As has been previously stated, no solid nutriment reaches any plant except as supplied to it in solution in water. What are the ultimate possibilities of growth in any crop is unknown to us, but it would seem as though they depended greatly upon the supply of water that can be absorbed, sufficient nutriment, of course, being provided.

Rye grass, upon irrigated fields richly fertilized, has grown at the rate of one inch per day, and repeated cuttings have been made at intervals of fourteen days, during a season of several months. Crops of grass upon irrigated fields of a total weight of more than 80 tons per acre have been reported by trustworthy English farmers in one season.

Irrigated grass fields in Italy support easily two head of fattening cattle per acre, every year, and have long done so. In hundreds of localities in European countries are irrigated meadows, which have borne grass without any sign of deterioration within the memory of the inhabitants, or the knowledge of readers of local histories, although the crop has been cut and removed every year during this indefinite period. Whether or not these immense yields could be further increased by more skillful management is not necessary to inquire. Those products are so far beyond the dreams of an American farmer, that they may well be considered fabulous. But there is no reason to doubt the facts. On the contrary, they should be used as a stimulus for us to adopt, whenever practicable, the methods by which these crops are produced.

The average product of grass upon our rich bottom lands, will not exceed two tons per acre, and upon uplands one ton per acre is a fair average yield. After a few years the best seeded of our meadows begin to deteriorate and run out.

A change of crop is made, and the meadows are once more seeded down, to run out again in a few years. The cause of the failure is the heats and drouths which follow the hay harvest, and which cause a cessation of growth until they are past. Beneath a temperature which would be genial and invigorating to plant growth with sufficient moisture, the grass dies for want of the substances that water would afford.

The farms of most European countries consist of but a few acres in extent.

This is notably so of Italy, Spain, France, Portugal, and Holland, where a farmer who owns and cultivates twenty-five acres is looked upon as a landholder of more wealth and importance, by his neighbors, than is the American farmer by his countrymen, who owns and tills his hundreds.

These small farms, when owned by the cultivator, are made so productive by every known method of high tillage, that they place the owner and his family in position of independence. Already the days of large farms in this country are numbered. Owners and purchasers of farms are fast finding out that one acre properly tilled is capable of returning in yield more of profit than three as usually worked, and are cutting up their present holdings, or making their purchases in farms of fewer acres. While this is so in all parts of the country, it is particularly so in the South, where the old plantations of hundreds of acres have made owners land-poor and brankrupt tax-paying.

The possibilities of the small farm here are as yet unknown, and when we tell our readers that five or ten acres, well cultivated, *and supplied with abundant water*, will yield, in the course of ten years, as much profit as fifty or a hundred acres, equally well cultivated, but without any provision for the necessary moisture, we are not overdrawing the fact.

Whether owning or purchasing, will it not pay the farmer to put the price of every other acre into an improvement which will give him command of the waters which nourish and assure his crops? In so doing it is within his power to increase production many fold, develop growth more perfectly, command a more ready market, secure better prices, and save labor and taxes.

To close this chapter, we copy the following from the pen of Prof. M. A. Curtis, of Jacksonville, Fla., editor of the *Florida Farmer and Fruit Grower*, it bearing strongly on the question of profit to be derived from Mr. Cole's system:

"Sub-Irrigated Gardens.

In the advertising of Florida as it has been carried on for a score of years, the practice has been to hold up the orange as the leading attraction, while market-gardening is made to figure as a secondary industry, a degree less genteel than orange growing, but a very convenient resource for meeting current expenses 'while the grove is coming into bearing.' However the other industrial resources of Florida may stand the test of experience, candor compels me to say that market-gardening in this State, as a rule, is as precarious an industry as can be engaged in. In the neighborhood of the leading winter resorts, where there is a good local demand, vegetable-growing may prove profitable, especially if no severe frosts occur in winter. Strawberries pay well where they are grown systematically, as at Lawtey, and good shipping facilities are at hand; but, as a rule, market-gardeners hardly recover expenses.

With natural waste, stealage, and high freight-rates, the net returns from market-gardening seldom exceed the expenses of packing and production. Other obstacles which are encountered alike by those who grow vegetables for marketing or home consumption, are found in the chances of severe frosts in winter and drouth in early spring, the latter amounting almost to a certainty. In the southern half of the peninsula danger from frost need hardly be taken into account, but in the northern half, winter gardens, which are the only ones planted for profit, are constantly in peril.

Drouths in Florida are particularly trying on cultivated plants whose feeding roots are within a foot or two of the surface. The soil in nearly all parts of the State is sandy, and lacks the capillarity by which soils of firm texture are enabled to withstand drouths. This evil is increased by tillage, which separates the coarse particles still more from each other, and it is lessened by admixture of such substances as lime and muck. On wild vegetation, growing on undisturbed land, drouth seems to inflict less damage than it does in other States. Without pausing to consider these seemingly contradictory phenomena, it suffices to say that the spring drouth in Florida is an obstacle that cannot be overcome or counteracted except by some system of irrigation.

In a soil so porous as that of Florida surface irrigation is impracticable. Windmills and other mechanical means of raising water are too expensive, except in peculiarly favorable locations. Artesian wells are most satisfactory, but they are not for poor men; and after a flow of water is obtained there arises the problem of distributing it to best advantage.

On the eastern coast of Florida, at Daytona, and a few other places, there has come into use within two or three years a system of sub-irrigation, which answers admirably for what may be termed intensive gardening, and in localities where there is good demand for garden produce. It is a modification of Mr. A. N. Cole's system—an adaptation of it to sandy land, the principle involved being the retention of water supply just beneath the roots, so that by capillary action it may readily be supplied to them as needed. The first essential is an abundant supply of water, as from an artesian well, and the second is a subterranean reservoir or trench which will retain a small quantity of water, and allow an excess, as from heavy rainfall, to escape.

The productiveness of the first gardens thus irrigated was so surprising that last winter a stock company was formed, with a capital of $50,000, for the establishment of an extensive garden at Ormond. In their construction parallel trenches are dug two hundred feet long by four feet wide and twenty-two inches deep, they being sixteen feet apart between sides and ends. A thin mortar is then prepared, composed of one part Portland cement and seven parts sand, which is poured over the bottom within a rough curbing of loose boards. After being smoothed to a level by drawing a short piece of timber over it, this bed is allowed three days to harden. Then a border is made all around by pouring the same kind of mortar between two courses of boards four inches wide, set on edge one inch apart, the space within the border measuring two hundred feet long by three feet wide. After the sides have been left three days longer to harden, the inner boards are removed and the whole inner surface receives a thin coat of pure cement, in order that it may be water-tight.

At either end of the cemented bottom are set wooden boxes six inches square, rising six inches above the surface of the ground. Through these the gardener may see how much water is in the trench, and regulate the supply accordingly. Next to these are laid lengthwise of the cemented bottom five courses of narrow inch boards. Crosswise of these a second layer of boards is put on closer together than the first. Then Palmetto leaves are laid over the boards and the trench is filled in with earth. The water from the well is led through the garden in two-inch pipes and distributed to the beds in half-inch pipes, which open through faucets in one of the boxes that lead to the bottom of each bed.

It is found that by keeping up the supply of water in the trenches the ground over and between them is kept sufficiently moist in the driest weather, and that a most luxuriant growth of vegetables can be obtained. In order to economize space, different kinds of vegetables are planted in alternate rows—for example, several rows of beets, radishes, and lettuce between two rows of cabbages. A close succession is kept up also, the intention being to have the ground constantly and fully occupied with the most profitable varieties. The cost of such beds is considerable, but it is found that one measuring four by fifty feet can be prepared for $15 or $20, and this affords a garden spot sixteen by sixty-six feet in area, on which may be grown a very large quantity of vegetables of quality and size not obtainable in any ordinary garden. An acre of ground thus prepared costs $1,000, but it is believed that it will produce to the value of $3,000 in a year. The length of the growing season in Florida is greatly in favor of this system, and it is not improbable that it will come into general use in localities where there is large demand for fresh vegetables in winter and spring. It is, of course, equally well adapted for flower gardens."

A. H. Curtiss.

Jacksonville, Fla.

In the above article this method is spoken of as being a modification of Mr. Cole's system. This is an error. It is his system pure and simple ; principle, method, and materials used all being covered by him.

CHAPTER X.

Mr. Cole's Home and Model Plot.

At Wellsville, Allegany County, N. Y., on the line of the Erie R. R., and in the Upper Genessee Valley, is the now far-famed "Home on the Hillside," where dwells and works Mr. A. N. Cole, the originator and demonstrator of this new system of agriculture.

Here is located his model, which now comprises about five acres, and here results of production have been accomplished, which astonish beholders, and exceed anything heretofore realized by any method of tillage. Here, in midsummer, can undoubtedly be seen a greater wealth of production than can be found on any equal area, not under glass, in America, and here have the benefits and results claimed for this system in the foregoing pages been accomplished and demonstrated.

The location is a sightly one, the house a solid, comfortable, and attractive home; but the lands surrounding it, in their original condition, could not well have been more unpromising or unproductive.

It is a strong, clay soil, underlaid by hardpan, and Allegany hardpan is proverbial for reaching the third rail in the fence. Situated on a hillside, his land was washed, gullied, and cut up by the rains and melting snows of winter, remaining wet and cold all through the spring, and until the summer sun baked it into crusts or lumps almost impossible to pulverize. A scanty growth of white daisies, Canada thistles, and coarse grass was the only crop it seemed capable of producing, even under the best methods of cultivation which could be applied to it. In the spring of 1882, Mr. Cole began quietly, and unknown to any outside of his own family, to construct, on the steepest and poorest of his land, a model of the system he had evolved by years of study, observation, and experiment. But a small plot was completed in this year, but the results from it equaled and surpassed Mr. Cole's most sanguine hopes and expectations. Eighteen hundred and eighty-three saw this little spot of ground producing vegetables and fruits largely in excess of the amount which could be used by Mr. Cole's family, and the surplus was put into the home market, where their size, beauty, and perfection attracted the attention of the public, and brought people in large numbers to see his work and ways. During the spring, summer, and fall of 1883 and 1884, the work of trenching, fitting, and setting small fruits was carried on as rapidly as means and competent operatives could be found with which to do the work. The end of the year 1884 found about two acres completed, not perfectly by any means, as some indifferent workmen had disobeyed orders in essential regards, and in others had done their work in a most careless way.

In the month of July, 1885, a formal introduction of the system was made to the world. On the 7th day of that month there gathered at Mr. Cole's home about one hundred of the leading farmers, professional men, and representatives of the press.

By these a careful, thorough, and exhaustive examination of soil, crop, plant, fruit, and system was made, and in every instance their comments and reports were flattering in the extreme to Mr. Cole, his work, and methods. Some of these reports we will now give, as they answer the double purpose of showing Mr. Cole's methods and the results. The reader will notice that in the preceding pages, not a claim is made for this system that is not supported by reported results made by men who had or have no interest in the system in a pecuniary point of view.

The opinion of Dr. I. P. Roberts, of Ithaca, expressed to Mr. Cole, was as follows: "Yes, Mr. Cole, you do all you claim to accomplish. You gather the waters into your reservoirs and pass them through the soil rather than leaving them to run riot along the surface. You transform this hitherto shunned and dreaded hardpan into soft, porous, productive, and best of soils to the depth of your trenches, thereby enabling the roots of vegetation to descend deeply into the earth. You remove the stone operating as obstructions and diseasing the roots of plants, putting the stone where they will do the most good. You provide against floods and drouths; and to a great extent, if not wholly, defeat the effects of frost."

Extract from the report appearing in the Buffalo *Express :*

"Yesterday was a notable day for Wellsville, or at least for one of its most widely-known inhabitants. Nestled away among the ragged Allegany hills, this snug town of about 4,000 people has in common with the rest of the world its ambitions and its celebrities. The particular industry that yesterday took a long stride towards popular recognition and favor is one growing up on the western hill of the town, and known as Coles's system of underground irrigation. Reckoned either as a freak or curiosity, or better than both of these, as a step into the next century in the domain of agriculture, this little plot of five acres of land, only two of which are as yet developed, will bear the closest inspection of either the skeptic or willing convert.

But perhaps everybody is not aware of the system, now under practical trial, which promises to revolutionize the world's agriculture—nay, according to its enthusiastic author, has already done so. Some four years ago Mr. A. N. Cole, better known as the 'father of the Republican party,' and the veteran editor of the Genessee Valley *Free Press*, began to put in operation a system of agriculture based on underground irrigation, an idea entirely his own. He had been studying the system a number of years before that time, but had not until then carried it into practice. There were drawbacks that need not be mentioned here ; and there was of course a town full of people who laughed at the idea as a crazy notion sure to come to nothing. But Mr. Cole persevered, and it is safe to say that yesterday he was able to demonstrate his success so entirely as to ensure him the title of the proudest man in Western New York, and perhaps out of it as well."

From beginning to end of a two-column article, commendatory throughout, the correspondent of *The Express* drew a faithful picture of what he saw on this occasion.

Report of Mr. Charles A. Green, a special correspondent of the N. Y. *Tribune*.

After a few preliminary remarks, Mr. Green said:

"Mr. Cole has been studying irrigation since he was seventeen years old, but his present system flashed upon him within the past few years. He has not yet extended his working model over more than two to three acres. I shall attempt to explain what I saw, and to state the claims of Mr. Cole as clearly as I can, considering our brief and frequently interrupted conversation.

We were first shown a patch of strawberries containing nearly two acres. These plants were grown in hills about eighteen inches apart each way, mulched with forest leaves, liberally fertilized with yard manure, and irrigated after the new method. I was told by the former owner of the hillside that when he sold it to Mr. Cole it was an unproductive piece of ground. The soil proper was not over ten or twelve inches deep and rested upon a tenacious, clayey hardpan—this was impervious to water. He said the frost acted so seriously upon this soil, on account of the surplus water not being able to escape through the subsoil, that it was almost impossible to keep plants alive in it during winter. Even the fence posts would be thrown out by the frosts in a very short time. A prominent contractor, who was walking by my side at the time, said that all that section of the country was underlaid by this peculiar subsoil, which is a great drawback to plant growth. I was also informed by this same gentleman that this part of Allegany County was not favorable for strawberry-growing, or other fruit except apples; and that the supply of small fruits is received largely from other sections. Mr. Cole has planted numerous varieties of strawberries upon his side-hill, among which I recognized the Bidwell, Sharpless, and other familiar varieties. The fruit was of an astonishing size, and grew in great abundance. While I live in a strawberry country, and am myself a strawberry grower, I cannot remember when I have seen so fine a display of strawberries growing upon the vines as I saw here. There were, however, evidences of high culture. A gentleman by my side echoed my sentiments by remarking that we could increase the size of fruits in our own gardens by such thorough cultivation as this. Adjoining the strawberries were growing different kinds of garden crops; also currants, raspberries, blackberries, potatoes, and a few fruit trees. At one point, where the ground was terraced, I noticed, growing on the rugged edge, a row of onions. I called attention to the fact that while these onions were on the very brink, there was no indication of their being disturbed by washing of the soil, as might have been expected in such a position. In fact, everything showed that in no place had the rainfall run down the surface as ordinarily, to the detriment of everything growing thereon, as the water passed into the drains underlying."

Report of Mr. R. S. Lewis, of the *Progressive Batavian:*

" Mr. Cole's farm, consisting of five acres of what was four years ago, and a part of which is now, a sterile hillside of clayey soil, so poor as to grudgingly yield sufficient substance to grow field daisies. It is as steep as the steepest part of Burleigh Hill Pavillion, the Bethany hill just east of the Center, or any other hill in Genesee County of which we have any knowledge ; and as to its ever becoming profitably productive, we don't believe there is a foot of land in all our county which was equally unpromising. Some thirty years ago Mr. Cole conceived the idea that plant life might be greatly, almost immeasurably, stimulated by underground irrigation. He had neither time nor opportunity then to perfect and test his thought, but it continued to simmer through him, and to recall itself to his attention again and again as the years passed on.

His conviction on the matter was greatly strengthened and stimulated by a conversation with Mr. Horace Greeley, in which that gentleman told him what he had heard of the wondrous productiveness near Los Angeles, Cal., where vegetation was fed by a subterranean river. Mr. Cole had thought and investigated until he had no doubt about the fact of a theory; but how to accomplish the irrigation—how to make his thought practical—was the question.

At last how to do it dawned suddenly upon him; the mists of questionings and doubts were gone; his dream of the years had materialized; his vision was clear. Where could he better test and demonstrate the truth and value of his discovery than on his own sterile, unpromising hillside? Along its eastern front runs a highway, with wayside gutter adjoining his land. Parallel with this, and some forty to fifty feet apart, and across about half his land to its highest boundary, he caused a series of trenches, about two and one-half feet wide by four and a half to five feet deep, to be dug, and filled to within eighteen inches of the surface with coarse large stone, covered with loose flat stone, for subterranean water reservoirs. These reservoirs were connected by numerous shallow and smaller trenches partly filled with small stones at about eighteen inches from the surface, and designed to carry off from trench to trench all surplus water. After the laying of the stone all the trenches, large and small, are covered with straw, or litter of any kind, as in ordinary ditching, and then covered with the soil again. Thus each large trench is a reservoir capable of holding from two to two and a half feet of water, through its entire length before it reaches the height where carried off by the cross trenches. The waters from the rains and melting snows, instead of passing off in surface rills and channels, is caught in these reservoirs and slowly and continuously filters through the soil from trench to trench—sweats through it, so to speak, rendering it porous, pliab'e, spongy—always sufficiently damp to feed and stimulate vegetation to the highest degree, and yet always sufficiently dry to be in the best possible order for cultivation.

On a part of his plantation which Mr. Cole has thus treated, he last year cut three crops of timothy grass, each crop being in the head when cut. Most of the trenched ground is now planted with blackberry and raspberry bushes and strawberry vines. What the berry bushes will do yet is only conjectural—they have a strong, healthy, prominent development—but the strawberry vines, it is utterly impossible to describe their wondrous wealth of productiveness. The vines are literally loaded with berries, and their average size is marvelous. Many were readily found which measured nearly eight inches in circumference, and there were no small berries. Mr. Cole proudly said : I have berries this year as large as peaches. He claimed he would harvest this year more bushels of strawberries from his vines than any farmer would grow bushels of potatoes from the same area of ground.

One or two facts more are worthy of mention. First. While the land all around this plot was frozen several feet deep last winter, this ground was not frozen—the plants grew the winter through.

2d. One of the deluging rains, so prevalent this season, poured down upon Wellsville a few days since, and while the hillsides all around were furrowed and ditched by the running waters, this plot was not washed in the least. The torrents sank into its porous soil, and were caught in its reservoirs, and the surplus passed off through its transverse trenches without in the least disturbing its surface or the crops grown thereon."

Extract from the report of Mr. James McCann, President, and Mr. G. W. Hoffman, ex-President of the Farmers' Club of Elmira. Mr. McCann, since President of New York State Agricultural Society,

The opening of the report being descriptive of the construction of the system, we omit that part.

"The soil is what I may call clay loam, with stones intermixed, but no appearance of sand, the close, compact subsoil not easily penetrated.

I refer to condition before treatment, and of this I had fair opportunity to observe in the adjoining land not yet brought under the new system; also in an excavation in progress where workmen had to strike heavy blows with their picks to penetrate the hard clay. The land treated by Mr. Cole was originally part of a considerable tract that was regarded as extremely poor, and my observations lead me to conclude that the estimate was just. The most striking effect of the treatment, as it seemed to me, was entire change of character, particularly mechanical condition, due, in large part, no doubt, to the very thorough manipulation, for it is not comprised in the trenching alone.

The entire area is dug up to the depth of fifteen inches, and all stones of any considerable size, even down to an inch in diameter, removed, thus changing mechanical conditions to such a degree that one is impressed with the great difference between the land treated and that immiately adjoining.

You step upon the trenched land anywhere and you find the soil yields to pressure of the feet, not a spot where it is not soft and yielding ; but on the land adjoining, it is hard, and the foot makes no impression whatever.

Another change is in color. That hard, forbidding clay, has taken the appearance of muck, or, at least, the color of muck and loam intermixed. Its texture is aptly described by Mr. Cole, who calls it an earth-sponge.

We were called to examine strawberries from plants set, as we were informed, last October, and I am free to say that the plat was a very interesting object, inviting study. There was a full crop of most remarkable berries—remarkable in size, color, and quality. I cannot undertake to estimate the yield, but it was certainly very large. I called Mr. McCann's attention to one plant of older setting that had ripe berries, and others in the various stages of growth, enough, I thought, to fill my hat if they could be picked at one time. One peculiarity of these berries was the absence of what may be termed a core, or hard stem in the middle; they were juicy and tender all the way through. As to foliage, I can only say that I never saw anything like it. I measured a leaf that was five and one-half inches across, and I plucked a broader one, with Mr. Cole's consent, and brought it home.

I must say that the changes wrought in the soil and its products constituted a great surprise.

As to the soil, I could judge by comparison with land that must have been originally of the same character. It now lies hard and compact adjoining the renovated earth, that, under Mr. Cole's treatment, has certainly become very fertile, whether with manure in abundant supply, or not, I am not prepared to say. The soil under treatment has the appearance of being thoroughly enriched with manure ; then there is the water supply for the roots to reach and use, obviating drouth apparently, and, besides, there is entire freedom from washing. Heavy showers had fallen in the week before our arrival, but there was not the slightest appearance of washing, and Mr. Cole informed us that all danger from washing was obviated ; a statement which I can accept as true, for he has provided reservoirs into which all surplus of water must pass, and if there is too much the overflow runs from one to another reservoir. Besides all this, the earth worked to find tilth serves as a sponge to take in a great deal of moisture and retain it for the use of the plants.

When I see a crop of strawberries much larger than I have ever seen under other conditions, no dead leaves, no runners, growth most luxuriant, and long succession in bearing, I must say that results are convincing. There were other proofs, about which I am not so well prepared to judge. For instance, an apple tree standing on this improved land was reported worthless, its fruit gnarled and valuless before the land was trenched, now bearing largely and fruit of fine quality. Of course, . I cannot say how much difference there is between the tree as it now appears and as it was before the land was improved. I observed, however, a young tree, the trunk five or six inches in diameter, perhaps, its growth most vigorous, the limbs smooth as if recently washed with lye, foliage fresh, full and green.

On inquiry I learned that it had only ordinary treatment, the limbs had not been washed, and its vigorous growth was attributed to the system of trenching and irrigating that increased the yield of strawberry plants and the size of fruit, the effect being visible in growth of all kinds."

There is no better place than right here to say that the change in the soil, noted by Mr. Hoffman, change in color, richness, texture, and all, was not the result of working it over, or of manures applied. Since the date of his visit, the trenches have been extended over lands where they have not had this thorough working, and no top dressings of manure have been applied, and yet these same changes have appeared in the soil, and the same results to vegetation.

Letter from the late Hon. John Swinburne, ex-Member of Congress, Mayor of Albany, and at one time Health Officer of the Port of New York. In professional life he stood in the front rank of American physicians and surgeons.

"Albany, May 7th, 1885.

Hon. A. N. Cole:

Dear Sir:—After quite thorough examination and consideration of your invention, or system, styled by you 'The New Agriculture,' I have become deeply interested in the matter, and beg leave by letter to express to you the impressions I have formed in reference to it.

Careful thought about the system impels me to the conclusion that as a plan for the storage and preservation of waters for irrigation, and purposes of general use, it demands and merits far more attention at the hands of farmers, gardeners, and the public generally than has as yet been given to it.

In a country like ours—in the eastern, southern, and central portions fast filling up with large cities, and villages, and thickly populated neighborhoods—the question of the most available means of obtaining a proper and sufficient supply of water for mechanical, manufacturing, and household purposes, and for protection against fires, is calling to its consideration the earnest attention and careful study of many of our ablest scientists and most practical thinkers ; while to agriculturists, manufacturers, and mill owners generally, in these sections, the very perceptible decrease in the volume of our rivers, creeks, and other irrigating streams upon the sufficiency of the supply of water from which they have been compelled heretofore (some in part and others wholly), to depend for success in their various avocations, has been to many of them the cause of great diminution of business, and business profits, and to others a subject of deepest anxiety.

The reduction of our forests, it is said (and very properly, too), has resulted in a consequent reduction of our rivers and streams, which were once freely navigable from their mouths nearly to their sources, until they are now only kept open for commerce, in many parts, by the application of great labor and large expenditures of money almost continually. As have failed these large streams, so have their smaller tributaries (from which they all, in fact, derive their supplies), become less in volume, until at length farms which were once properly and abundantly watered, are now comparatively without supply, and streams which once furnished sufficient water-power for the running of mills and factories, now scarcely afford power sufficient to propel the churns of farmers occupying their banks. The depreciation in the value of lands in many parts of the country for agricultural purposes, and the supply of crops therefrom and from the same cause, has become equally perceptible. Yet, the supply of water from the clouds, from rains and snows, has not, so far as we know, in any way decreased ; but the forests are not here to husband them, and these waters are permitted to soak into the ground, or run to waste from the surface almost as soon as they strike the earth.

The problem heretofore has been how best to secure and husband these supplies by artificial means, so as to most effectually preserve them for the vast demands of our wonderfully increasing population, for family and business purposes, and especially so as to make them more useful in the cultivation of the soil.

Many able and ingenious thinking men have for a long time given this question their attention ; and many plans have been suggested— some of greater and some of less merit, but all accompanied with an apparent intricacy of detail, and weight of expense in their application, which has prevented the general or considerable adoption of either.

But you, Mr. Cole, seemed at last to have discovered a scheme, plain and practical in itself, and evidently of but moderate expense in its adoption to the uses and necessities of a very large proportion of the people who are now suffering severely from the evils to which I have above called attention. You style your system ‘The New Agriculture,’ and from its probable effect upon agricultural districts in which it may be hereafter adopted, as indicated by the experiments you have already made, the name would not seem to be in any way misapplied. If the result of its use in general should be an increase in crops and vegetation, to but half the extent foreshadowed or promised by those experiments (and I can see no sufficient reason why your claims in this respect may not be fully verified by practical application of your plan), you have developed and now offered to the country and the race a new system for husbanding the falling waters, and a new plan for their use which will not only establish a new era in agriculture, but which may be so used as to afford the needed supply of good, healthful, and pure water for the other ordinary uses of life to very many sections of the earth, where the inhabitants are now suffering disadvantages, and privations from its want.

Your plan of invention is exceedingly simple in detail, and the greatest wonder to any one who shall see or read of it will be that it had not been thought of, developed, and adopted long before. It bears the impress of reason and sound sense upon its first presentation to the mind, and more mature reflection upon its merits only results in more strongly developing these characteristics in it. The scientist and the plow-boy alike can each, with equal promptness and facility, perceive its scheme and merits at a glance; and the person who proposes to use it on his farm or garden, or in connection with his shop, dwelling-house, mill, or factory will not require the assistance of the scientific and mathematical knowledge of the civil engineer or architect to enable him to put it in successful operation, the brains of an astute accountant to estimate its cost, or the eye or mind of the learned student of nature to discover its results. Combining in itself a plan for the accomplishment of these objects highly essential to the comfort, convenience, and business interests of the people—storage of water, irrigation, and drainage—it will be seen at once, by even the ordinary mind, upon the most casual inspection, to be practicable and feasible for either purpose ; and it must be equally evident that great advantages must accrue to the user of the system, either for agricultural purposes, the storage of water for other general uses, or as a means of drainage simply.

Scarcely a township exists in our country in which there are not many farms upon which your admirable system could be applied to great advantage and profit. Large portions of territory in agricultural districts are now entirely useless, or at least comparatively unproductive, by reason of insufficient irrigation, and these, through the appliance of your ' New Agriculture,' could be made vastly more productive ; while the present productive portions would be increased in productive power through the same instrumentality. To the grape and other fruit growers it seems to me it affords especial inducement for use, which will speedily bring it into imperative demand with the large majority of this important business class. Through it thousands of agriculturists, in every State, may easily, and with little expense, make their barren wastes to smile with productiveness, and the better portions of their farms to double in value by reason of increase in crops.

But the advantages from the use of your plan in storage of water for other than agricultural purposes are equally apparent, and must eventually bring it into active demand and use in localities where the supply of water is now insufficient for the requirements of cities and villages ; and by its application many such corporations will be enabled to furnish their citizens with good, cool, and pure water in sufficient quantities, and at far less expense than they can by any other plan or system now known. Of course, whether it can be so utilized as to furnish very large cities with sufficient supply is a problem hereafter to be demonstrated ; but in our own State (and without doubt in every other State) there are hundreds of small cities and thickly populated villages and hamlets, whose inhabitants are now suffering great inconvenience, and incurring risks of sickness and death from malarial and epidemic diseases from insufficient supplies of healthful and pure water, whose surroundings are such that, by the reasonable application of your simple system for collection and storage of water, they could each, at much less cost than in any other way, be furnished with a permanent and sufficient quantity of the best of water for all the purposes for which it may be required by them. Then, too, the hills or mountains surrounding or adjoining these places, often now utterly unproductive, and sometimes even unsightly in appearance, can, by this same plan, be transformed into productive and ornate terraced gardens, far excelling in products and profit the ordinary agricultural lands of the neighborhood, and rivaling in beauty the most famous of the ornamental gardens of the old world, presenting at all times a 'thing of beauty' to the eye, season by season, affording more profitable remuneration to their owners from the crops, and fruits, and vegetables which shall spring from and adorn their slopes; and at the same time and always affording to the inhabitants of the populous places beneath their shades a bountiful supply of Heaven's best and only beverage for man.

I am confident that your system will grow in popularity with its use; and eventually a grateful people, thankful for the blessings your invention has brought to their hands, will rank you as a benefactor of the human race, who has not only succeeded in making two blades of grass grow where one was wont to appear; but who has also taught them by simple method, and at cheapest cost, the way to secure for themselves a sufficiency of one of the most important of God's gifts to man, and beast, and nature.

This letter requires no answer; it is written in testifying appreciation of the merits of the invention of an old friend, and he is at liberty to use it as he may deem proper.

With sentiments of respect, I am, as ever, yours truly,

JOHN SWINBURNE."

Mr. Cole's present residence was built by Mr. Wm. Pooler still a resident of Wellsville, and the farm, which contains about 50 acres, was owned by him. Visiting the place and noting the change in soil and production since Mr. Cole perfected his system, he sent the following communication:

"I think it was about 1850 that I purchased the place on which you now live. The hillside had been cleared for several years, being one of

the earliest lots improved in what is now Wellsville, then the town of
Scio.   There was an old orchard on the place, and also a tree, which I
shall never forget.   It was not in the orchard, but stood by itself, a little
to the north-west of the house and was a Roxbury russet; no more worthless fruit could have been anywhere found.   Yesterday, September 22,
1885, I plucked from this tree two apples; one, the smallest I could find,
the other of average size of those of which the tree was so loaded as to
bow its branches to the ground upon which the lower limbs rested.   I
should judge there were twenty-five bushels of apples on this tree, two-
thirds grown.   These apples, on the first of October in the years 1853,
1854, and 1855 did not average larger than crab apples at that time of
year.   They were not so large at harvesting as the small one I picked
yesterday, nor were they quarter as large as most of the apples on the
tree at this time.   The tree was then about ten years old, and was covered with moss, and in all respects of no value, and I threatened, at the
time, to cut it down as a cumberer of the ground.   I should guess that
the tree might possibly have borne two bushels of apples in a bearing
year, and we did not pretend to gather them.

The apples now on the tree are large, fine, and fair; in fact, they
are the finest russets I ever saw.

You showed me early rose potatoes, grown this year, the like of
which I never saw anywhere.   Some of these weighed from a pound to a
pound and a half apiece, and I should think one would weigh two pounds.
You assured me that you had grown them at the rate of over one thousand bushels to the acre the present season, and I have no reason to
doubt it.   As there is no fungus on your grounds there is no rot.   The
tomatoes, all over the town, are rotting, but I did not observe any rotten
ones on your place, and I certainly never saw such splendid fruit, nor anything like as many to the plant.

I gave you an account of my experiment with two acres of potatoes in 1854, and here repeat it.   The plot in which I planted is a portion of the ground now embraced in your garden, on which this year has
been grown such crops as I never set eyes on before.   I fitted these two
acres with greater care and painstaking than any equal amount of ground
in my life, mixing a portion of the subsoil with the surface, and covered
it deep with well-rotted barnyard manure, making it very rich.   A careful man and a good farmer planted the two acres to potatoes on halves,
and I realized just thirty bushels for my half.   This completely discour-

aged me, and though there was no better house in Wellsville than the one I had built upon the place, and the barn was nearly new, I gave up, and sold the property for what I could get.   You told me yesterday that you valued your two acres completed at $5,000 an acre, and that it was paying well at that.   As $5,000, at 6 per cent. interest, only gives $300, I do not wonder; since I am sure you are getting two or three times that from an acre.   I have seen these strawberries and other fruits and vegetables as sold in this market for the last three or four years, and have eaten of the fruit, and have never seen anything anywhere near as large, beautiful, and fine flavored.

You yesterday showed me pods of peas, and I carried home specimens with eight peas in a pod, of such marvelous size as to astonish me. The peas were of the dwarf variety, as shown by the vines, and yet they were as large as Delaware grapes.   You assured me you grew five hundred bushels of pods to the acre of these peas, and I believe you, since your Champion of England, on vines higher than any man's head, loaded with pods, and still covered with blossoms, presented such a sight as I never saw before.   Your squashes, beets, cabbage, and cauliflower were all very fine; and as for squashes, I never saw anything in my life so astonishing.   Though quinces are rarely grown in Allegany County, I saw as fine ones as I ever came across anywhere.

Nothing so much surprised me as the change wrought in the soil. The cold clay and hardpan had been turned into a soil, deep, soft, and very rich, growing all forms of plants, bushes, and trees to perfection. You say your system wipes out the hardpan, and it certainly does.

This latter feature of your plan surprises me more than any other, but perhaps I should except from this your spring brook, and that stream of pure cold water flowing out from the pipe in the rear of the house, there being no springs on this part of the place.   Nobody can look down into your trenches when open, and see the long stretches of spring water in them as I did, and not discover that you save all the water which falls upon the hillside, using what is needed for the growing crop, and the remainder, by far the greater portion, running off in purity.   Though before my visit of yesterday I was convinced your system was a success, I left your place prepared to say what I now do.

Your discovery has no equal, nor do I believe anything will hereafter be discovered so important to health and prosperity of the people.

WILLIAM POOLER."

In a letter bearing date, March 28th, 1885, addressed to Mr. Cole, the Hon. C. R. Early, one of the most celebrated among physicians of the Keystone State, and none more eminent to be found anywhere in knowledge of causes and effects of fungoid infections, the Doctor, among other things, says :

"You may remember, in the last of June, 1883, while I was in Wellsville, I was invited to dine at your 'Home on the Hillside,' in company with several other parties interested in developing the oil, gas, and timber resources of Northern Pennsylvania, and Western New York.

At your table we were much astonished to see the most delicious fresh peas, just picked from the vines, and the finest strawberries that all acknowledged ever having seen. The question was asked :

'Where do you get such fine peas?' Your answer was: 'They were picked from my own garden.'

'Where do you get such berries?'

'They are also picked from my own garden.'

'Come, now, Mr. Cole, that will never do. I was raised in Allegany County. This is too early in the season for either peas or strawberries. Besides, Allegany never produced such peas and berries as these.'

Your reply was, that this was the fruit of your system of underground irrigation. You then explained to us your system of sinking troughs in the ground, and taking up the water as it fell, and holding it back to supply moisture to vegetation as it was required. This was entirely a new feature to us all ; and, after dinner, we repaired to your garden, a lot on the hillside, where you explained to us your system in detail. The more I examined the more I was astonished to find every bush, twig, stalk, tree, and fruit perfectly clean and healthy. No rust or fungi of any kind whatever was to be found. You showed us a stream of water coming from the trenches, a continuous, bright, and sparkling brook, and yet it was a dry time ; quite a drouth. But in spite of all this, we found a stream of water coming from your hillside constantly, with no spring to feed it—only coming from the stored-up rains and dews that fell, caught up and garnered by these troughs, furnishing a constant vapor to the roots of your vegetables and plants, keeping them in uniform condition of moisture ; never too wet, never too dry. This system made a very deep impression upon me, and upon returning home and thinking the matter over, you will remember I wrote you a letter, suggesting that by the use of natural gas (which must take the place of coal and wood for heating purposes) to heat the water in the Fall and Spring, and running steam pipes through the troughs (or dropping the warm water into them) to keep the water warm, you might raise all kinds of produce, and, as it were, do away with Winter. You could do, as I found while in Europe was done there—produce the finest pineapples by use of this warm-water system, thus doing away with expensive hot-houses. In this letter I also suggested, that, where you wished to raise tropical fruits, you could throw a canvass over the space to keep off the winds and snows. All this was in the most part a joke, as applied to Allegany, but upon receiving your reply, I was astounded to read that you had already obtained a patent covering these points.

On the second day of July last, I was again in Wellsville. In passing through the streets I noticed on the corner baskets of strawberries; some were small, diseased-looking berries, but alongside of them were luscious ones, nearly as large as peaches. Said I:

'How much are your strawberries?'

'These are sold at thirteen cents, and these at twenty-five a quart,' was the answer made by the vender.

'But why should there be such a difference in price?' I inquired.

'Why! these are Cole's berries.'

'Cole's berries! What do you mean by Cole's berries?'

'Why! they are raised here in town by Mr. Cole.'

'Who is Mr. Cole?'

'What! don't you know A. N. Cole?'

'Oh! yes; he has been termed the father of the Republican party. So he raised these berries here in town? Well, I do know A. N. Cole, and I think he has succeeded in raising better strawberries than children, for let me tell you that his Republican children have given us Democrats a mighty sight of trouble.'

'Yes, sir; that is so. He generally succeeds with anything he undertakes.'

'How many of those berries does it take to make a quart?'

'About twenty to thirty; I suppose an average of twenty-five would cover it.'

'Why! You had better sell them for a cent apiece.'

'Well, they sell at that as fast as lightning. They don't stay on hand long.'

Of course I knew whose berries they were as soon as I saw them. It was only a whim of mine to interview the groceryman.

The next day a party of us were visiting your grounds, and you may well remember the liberties taken by me at that time. I then had with me a powerful glass, and I was determined to investigate matters thoroughly. I examined the roots, leaves, stalks, and berries of your strawberry vines. I dissected and investigated them in every imaginable way, as also the pea vines, cauliflowers, cabbages, and in fact all vegetables and vegetation within your grounds, and, as I told you at that time, I did not find a single exception wherein a plant was not perfectly clean and healthy.

No fungi to be found anywhere. Root, stalk, leaf, twig, and fruit, all in perfect health, and absolutely free from fungus or parasites. Strawberries larger than plums, and everything in like proportion. Even the timothy and other grasses seemed brighter, fresher, and more luxuriant.

Of course, I inquired into the expense per acre of such a system, which I cannot pretend to give from memory; I will leave that to you. But allow me to say that if all lands were, in place of underdraining and subsoiling, treated as you do yours; deep trenches, broad and wide, filled with stones and covered with soil as yours are; fertilized with a compost as you prepare it, having it fully assimilated before using—I say, if all this could be done, then, and not till then, can we do away with this creation, cultivation, and dissemination of poisonous fungi, which, as has been shown, is working such sad disaster and death to the whole universal and vegetable world."

CHAPTER XI

That the Department of Agriculture among Cabinet positions is to be found most important of any in the future of our government may be set down as a fact. The enactment into law of what is known as "The Hatch Experimental Station Bill," the author of which is the Hon. W. H. Hatch, of Hannibal, Mo., dated an epoch with the people of the United States, to ultimate in such an education of the industrial classes as to place America in the front rank of progress, and make the entire country one of universal prosperity and wealth. At this point I would ask the Department of Agriculture to take means at once to lay before the American public that portion of a report on irrigation in the United States, made by Mr. Richard J. Hinton, under direction of Commissioner Coleman, found on pages 150 to 154, that our people may realize the fact, the saving and use of the waters for all they are worth is to become the world's future. I especially ask that all agriculturists and horticulturists may note the conclusions reached by Mr. Marsh, late Minister to Italy, touching the duties and responsibilities of governments in taking charge of the whole matter of irrigation. To this complexion must it come at last. That this great revolution is at our very doors, let the following from *The Field and Farm*, of Denver, Col., bear evidence, that the policies of government have already been substantially fixed as regards this whole matter:

"IRRIGATING THE PLAINS.

Last year Congress appropriated $250,000 for the purpose of investigating the extent to which the 'arid regions' of the West can be redeemed by irrigation. It is proposed to dam up the canons of the Rocky Mountains so as to form immense reservoirs of water from the melting snows and heavy rainfalls of the region, to be used for the irrigation of the arid lands west of the 100th meredian, embracing an area of 150,000 square miles. The location of the reservoirs, and the selection of the courses of irrigating canals are under direction of the originator of the gigantic scheme, Major Powell, of the geological survey, and will ultimately cost between three and four million dollars, while the construction of the works must entail an outlay of hundreds of millions of dollars ; but, in return, it is estimated that the full realization of the plan will reclaim for profitable agriculture an area equal to four-fifths of the present cultivated land in the United States. The first appropriation was made chiefly to obtain more data for determining whether the scheme was so feasible as to justify further expenditure. Congress has now virtually committed the country to the execution of the project by making another appropriation of $250,000 for continuing the survey. The governors of all the Territories to be first and chiefly benefited by the project speak loudly in favor of at least the present work, and urge local co-operation."

In view of the fact that great artificial reservoirs are a constant menace, likely to give way at any time with disastrous results, we ask that the merits of this system be given consideration by those in authority. Under it the waters of the mountain streams can be turned into underground storage, and carried from points along summits to slopes and inclines, foothills, plains, and into valleys, until moisture shall everywhere prevail, and the arid lands become a water preserve down to the bed rock. That such a system would be automatic and at all times operative, cannot fail to be discovered by the most casual thinker. Of the efficacy of Mr. Cole's system and the profits realized, it is only necessary to compare them with the system of intense culture applied by the market gardeners in the country districts around Paris, as reported by Prince Krapotkine, results electrifying the world.

He asserts that there is not a country in the world that begins to contain the number of people that might easily be supported upon its own soil without importation of food or agricultural supplies of any kind from foreign countries ; that where there has been an assumed pressure of population, as in England and Ireland, the causes have been, not that the land was not abundantly able to procure for them all, and more than all, that was needed for a full and satisfactory support of existence, but because in these countries the land laws and the monopoly of land by private ownership have been such that the people have not been able to utilize its resources as they should, and, under other circumstances, probably would.

As an illustration of the productive capacity of the earth under proper treatment, he gives a number of instances borrowed from the experience of the market gardeners in the country districts around Paris, where the soil, even in the hands of relatively ignorant men, has been utilized so as to be enormously productive. He refers to one farm of 2.7 acres in extent, from which there are annually taken 125 tons of market vegetables of all kinds. The farmer in this case—and he is but a sample of his class—has found out a part of the secrets of nature, and by the building of walls to protect his lands from the cold winds, by whitening these so as to secure all possible radiated heat, and by a constant and judicious use of fertilizers, has his little farm constantly in a productive condition from January 1 to December 31. He has, in effect, by simple and inexpensive means, produced a result equivalent to what would have been obtained if the farm had been located in the open air, a number of degrees to the south of Paris.

He and other market gardeners around Paris make their soil, and have each year quite a quantity to sell; for, if this disposition was not made, in consequence of the amount of fertilizer used, their farms would gradually be lifted up above the level of the surrounding ground. Prince Krapotkine says it does not in the least matter what the soil is from which they originally start; for a French market gardener would in two years' time raise an abundance of vegetable products from an asphalt pavement as a foundation. Soil, either made up of loam or fertilizers, is a chemical product, and not the least difficulty will be experienced, when the laws of chemistry are better understood, in manufacturing all of the material needed for plant life.

In his opinion the ordinary French market gardeners are but beginning in the business, for they devote a great deal of unnecessary time and labor to work which could be much more easily performed by mechanical processes. The way in which this could be brought about is shown in the experience of one French market gardener, who has covered over a half-acre tract of ground with a glass roof, and has run steam pipes, supplied by a small steam boiler, at intervals under the ground sheltered by this covering. The result has been that he has been able for ten months out of every year to cut each day from this little tract of ground from 1,000 to 1,200 large bunches of asparagus, an amount of product which, under ordinary conditions, would require not less than sixty acres of land; that is, by a skillful adaptation of means to ends, the productive capacity of a given tract of land has been increased more than a hundredfold. Even this result, great as it is, has been surpassed by an English gardener, whose experiences are referred to in the last number of the *Quarterly Review*, who has entered into the cultivation of mushrooms, and has been able, in consequence of the skill he has shown in this work, to net an annual profit from a little farm of one acre in extent of more than $5,000. That the market gardens around Paris are profitable may be judged by the fact that the average rental charged for these is $150 per acre.

Prince Krapotkine maintains that, even at the present time, with their only partially instructive methods, the French gardeners could easily raise enough, both in animals and vegetables, to supply all that was needed for the sustenance and protection of life at the rate of 1,000 human beings to the square mile; or, in other words, under a method of intense and properly directed culture, it will be easy for the State of Massachusetts to sustain within her own borders a population of not less than 9,000,000 human beings, and this, be it remembered, is but the beginning, for no one yet knows the limit to be set upon the productive capacity of the soil.

To create soils, infuse, and continue within them elements of fertility, as is done and being done by these French market gardeners, would call for an outlay in the complete regeneration of France alone of greater amount than it would to cover the watersheds of the world under Mr. Cole's system, accomplishing the same results, and, in addition thereto, saving and giving for use to man and crop the waters.

Here we will give place to selections from a letter written by Mr. Cole to Mr. Farrar, of the *Labor Review*, and by him published, as he, with his ready pen, tells of results of the system which will follow its introduction—results not undetermined, but proven by Mr. Cole and others who have adopted it.

[From the Labor Review.]

"The following coming to our table from Father Cole, of the Home on the Hillside, will need no extended introduction. If there is now living anywhere a man appropriately denominated the Father of the Republican party, that man is Asahel Nicolas Cole, of Wellsville. For years, this bold, brave, fearless, and independent Old Man of the Mountains, as he is denominated, has made no concealment of the fact that he is a labor and land reformer, and in sympathy with every endeavor of the industrial classes to make their way up in the world. This man has been from the first little understood, scarcely understood at all, in fact; at times declared by the Democrats a Democrat and free trader; again by

labor journals a labor reform man, and we believe also that the Prohi-
bitionists have claimed that he belonged to them. We have known in-
timately the man since so widely spoken of on account of his New Agriculture, sometimes denominated Aquaculture, or Sub-Irrigation, for many
years. But we have said enough. Hear Father Cole:

'WELLSVILLE, N. Y., July 21.—I made to you a promise somewhat in
pleasantry, and, having done so, making it a rule never to make good
promises and not keep them, and having made bad ones not to carry them
out, I proceed to write. I did not say what I declared could be done for
thirty dollars an acre, but, as I understand your soils, I conclude those on
the surface are naturally rich in mould, and have clay subsoils, and will
presume such is the case. Again, I conclude that stones are abundant
in your section, a matter of great moment. In its absence, resort to tile
is a necessity. Even in countries where recourse must be had to quarries,
the expense of fitting lands will not be greatly increased. I will say, therefore, right along, that in lands where round and flat stones prevail, these
are invaluable, since, once fitted, with the reservoirs and overflows such as
I am about to describe, there will be little work for you and your children
and children's children for thousands of years, except to sow your grain,
put in your potatoes and corn, harvest your crops of grain, grasses, fruit,
etc., and be happy. You will never have any subsoiling to do, and, as
for plowing, dragging, and working your land generally, inclusive of manuring, the rains and melting snows will do fully nine-tenths of it, and the
automatic inmovement of the waters will convince you within three years
from the time of fitting your first acre that there was a little more of
philosophy in the heads and hands of your New England ancestors, with
their stone in one end of the bag and the grist in the other, than in those
of their sons still holding on in the ways of farming such as have pre-
vailed the world over outside of the semi-barborous countries where folks
like the Japs, the Heathen Chinee, and some others have crudely done
for thousands of years what I am doing on my hillside.'

(Here follows description of manner of construction not necessary
to repeat.)

'You will now, I feel sure, understand me when I say the cost of agricultural or ordinary farm lands is placed at $30 per acre, and most per-
fectly fitted garden lands at $300 per acre. My most perfect lands, a
hopeless clay and clod, filled with stone (more stone than soil when I began), are a perfect tilth or sponge, made so by the fork to begin with, and

rendered absolutely perfect by movement of the water through them for five years, animalcule left dead in track of the waters, and are this hour the most productive soils on the face of the earth. On these I am growing cauliflower from the embryo of the size of a walnut to thirty inches in circumference in from seven to ten days, rapidity of growth depending on atmospheric conditions; the more hot and dry the weather the more rapid will the roots make for the water beneath. It is this that gives me a cauliflower, full grown, in a fourth of the time from appearance of embryo under ordinary conditions of growth.

Astonishing Results.

To show what Mother Earth is capable of doing at her best, has been my endeavor. Thousands of people have been here to see my work. When asked what it has cost my answer has been : " My very best acre may have cost four or five hundred dollars." In this I have included not merely the cost of fitting, but of planting to fruit and all else connected with the work for the first three years, inclusive of caring for and harvesting of crops. My most perfectly fitted lands have been tests simply, and on these I have grown early rose potatoes at the rate of 1,200 bushels per acre ; timothy and clover, three and four crops annually on the same test meadow, at the rate of three and a half tons of cured hay, aggregating from twelve to fourteen tons to the acre. I grow, year in and year out, at the best, strawberries at the rate of from four to five hundred bushels to the acre, and these from the size of ordinary ones to that of peaches, and thence on up in some instances to the size of the average greening apple. I have now growing on my place Cuthbert raspberries at rate of not less than six hundred bushels to the acre. My raspberries and blackberries are growing upon lands costing not to exceed fifty dollars an acre for fitting. I use no phosphates, nor would I take as a gift all that might be offered me. I use well-rotted barn manure, but where manures are deposited in open trenches, that these may be liquidated by the rains, there is no need of rotting. The animalculæ left dead in soils is my chief reliance for fertilizing. The value of waters falling upon lands during the year in this regard is far greater than all possible manuring. In passing the waters through the clay and clod from trench to trench, as I am doing, the effect is, to aerate, infiltrate, and render them warm, vital, soft, porous, and productive to the depth of trenching. And on lands where only three, or four, or five inches of surface of productive soil naturally exists, the roots of plants will, within three years after lands are trenched, penetrate to depth of the trenches. Best of all, the soluble elements of the clay and clod are released, and aluminum, potassium, and all else making up best of "soup, porridge, or broth," pabulum perfectly prepared for absolute perfection of plant growth follows.

The Poorest Soil Will Do.

More gradual will be found the fertilizing of arid and sandy lands, but none the less sure. New England, were her stone buried in her soils as I am doing, would, I am confident, within a decade be producing more of value than now grown on all lands under cultivation in the American Union. To drop the waters of rains and melting ices and snows into trenches, enabling them to descend to the bed rock, would fill subsoils with water, and, diffusing moisture in all directions, deserts would disappear, and in their waste places would grow grasses and grains, fruits and flowers, grand trees, deep rooting and sending up great trunks with spreading branches; and along mountain sides and in valleys, across arid plains of alkaline deposits, where the prickly pear, the sage brush, and grease wood now prevail would appear the higher forms of plant life, and springs everywhere cropping out, brooks would be grown, rivers appear in valleys where none now are found, and lakes of crystal water, alive with brook trout and other of the best varieties of fish would be found in dells among the mountains, and from these the waters can be dropped from mains to the valleys, and by use of the water motor and the turbine wheel with cable attachments the world can be set in automatic play of whiz and whirl, moving the most ponderous of of machinery, and setting the pick, and spade, and shovel at work, guided by the hand of man, the forces of nature made to do the work now done by hand, from that upon the Panama Canal, down to that of the housewife with her churn, the farmer with his threshing machine, the manufacturer with his mighty shaft in perpetual motion, and the lady with her sewing machine, driven by a motor hung upon the wall as an ornament, magically moving her knitting needles in automatic and endless movements.

Its Sanitary Advantages.

To conform soils as I have done and am doing would produce not only all of these results, but would drain all swamps, dry up all fetid pools and morasses, purify all waters, render healthful and vivifying the air, make an end of stagnation, living germs, microbes, malarial infections of air and water, and, making an end correspondingly of premature decay and untimely death ; so would the harvest be. To calculate the effect upon the climates of different countries is a something in the contemplation of which the mind of man can scarcely embrace in its utmost stretch of endeavor. To diffuse moisture through all of the earth's watersheds from pole to pole, as I have demonstrated, is as easy as to sink trenches on your own farm, following directions herein given, and would have the effect to everywhere maintain the dew point in soils, winter and summer alike, simoons, siroccos, cyclones and blizzards would disappear, and the winter of earth's discontent turn to glorious summer.' "

Reports from others who are using the system are as strongly in its favor as those from Mr. Root and from Florida. We have not the space in which to publish some we would like to give our readers, containing as they do information of value. As the question has been frequently asked of effect on orchards, we will give room to a letter, or an extract from a letter from Mr. Cole, and a report made by a gentleman who has applied it to his orchard. Mr. Cole says : " Before the war I visited the farm of Mr. Christopher Grastorf, an intelligent German farmer of this town. On it was found a remarkably thrifty and growing young orchard of four acres. Immediately above the orchard is a rocky cliff, at the base of which issued a large number of living springs. After perfecting my system, I suggested to my friend the advisability of sinking a trench below the cliff, and throwing the water in even diffusion along the water-table, insuring sub-irrigation the year round. No disciple could have been of readier agreement, and he forthwith followed the suggestion. Results were simply incredible ; grass, patatoes, garden vegetables, etc., were grown from year to year between the trees, and such crops have never been seen within my knowledge, outside of my 'Home on the Hillside.' This, the 6th day of June, 1889, Mr. Grastorf brings to my home specimens of the northern spy apple in as perfect condition as were any fruit from any orchard of last fall. These he says are but samples of the yield of this orchard, and that he is not able to thus preserve apples grown without sub-irrigation."

For himself, Mr. Grastorf says : " I became convinced that one acre of sub-irrigated land is of intrinsically greater value than ten or twenty under old methods of culture, and decided to remove my barns, at an expense of one hundred dollars, that the liquids from the barnyard might enter the trenches. This I did with best and astonishing results. The net profits from my four acres being much more than from entire fifty acres outside, and had I the means I would sub-irrigate my whole farm."

Office of Opera House,
S. F. Hanks, Agent,
Wellsville, N. Y., July 1st, 1889.

A. P. Cole,

Dear Sir :—Your father requested me to send you a statement of. my trenching for water in my garden lot. You know it is across the street from yours, not directly opposite, but northerly therefrom. On the edge of the upper slope I dug a trench five feet deep, running the length of the lot on Highland street. It was made to descend from each end towards the center ; at the center I laid a pipe, *i. e.*, at right angles leading down the hillside to my lane. This pipe was also buried below the frost line. I put a goose-neck pipe at its terminus at the lane, and have an artificial spring, as a result, of the purest water. I prepared the trench by filling the bottom eight inches with gravel and sand for the purpose of filtering and purifying the water.

I then filled the trench with stone picked from the garden to within eighteen inches of the surface, covered these small stones with large flat ones, and buried the entire length with garden soil. I cultivate over the entire system with astonishing results. This season has been very rainy, and the surplus waterfall on my lot has been conducted off by the underground pipe, leaving the surface ground under tillage light, spongy, and in a proper condition to promote rapid growth of vegetation. There is a large stream of surplus water running down State street on both sides, one from your father's, which is utilized by some German families living by the side of its course. Yours truly,

S. F. Hanks."

Mr. Richard Jacobs, of Independence, Allegany County, an old veteran of the late war, writes that he has secured the blessing of a flowing fountain of pure and sweet water for family use, and that at a point where before putting in the system there was no water supply.

In a letter of recent date, Dr. W. M. DeHart, of Stark, Fla., writes : " We are suffering from the effects of a seven weeks' drouth. Vegetation

is burned up. On my trenched land Irish and sweet potatoes are growing strong, green, and of fine color." Dr. DeHart is extending his work.

Space will permit of no more reports.

The year 1889 will pass into history as the year of floods, carrying
in their path death to thousands, and destruction to property and crops
too large to enumerate.

The "Home on the Hillside" was in the path of the storm, and
stands to-day unmarked and without damage by the falling, rushing
waters, while on all sides of it lands were gullied and washed, crops
destroyed, and other damage wrought. Though the fall of water was
such as was never before known in Western New York, the trenches on
Mr. Cole's land took up the waters, holding them in storage. Had the
watersheds of the floaded districts been under Mr. Cole's system, the
rainfall would have been taken up and held as it is by forest districts,
the stored waters feeding crop, spring, well, brook, and river in long and
steady flow. Floods can be, should be, rendered things of the past,
while our springs and streams are restored to their original purity and
volume. By these floods millions of dollars have been lost; other millions must be spent to repair damages.

A fraction of these sums spent by government or individuals would
have provided safe storage for all these rainfalls, making them agents of
good, and preventing them from becoming a demon of destruction. With
results of these storms still in full view, our readers will please peruse the
following eloquent appeal made to the tile makers of Ohio in convention
assembled at Columbus, three years ago.

The speaker was Mr. L. W. Matthewson, civil engineer, of Cincinnati, as reported for and published in *The Drainage and Farm Journal*
of Indianapolis, Indiana. Let every reader peruse and ponder, and
make up his or her mind whether aquaculture or the new agriculture is
a chimera, or, on the contrary, the greatest and most triumphant exhibit
of physical fact brought out since time began:

"The profits of farming by established methods are small, hence the
reluctance of farmers to take risks. They are impelled to close economy,
and it becomes a habit verging upon parsimony too often. But here is
at least a way of securing enormous gains, and the plan is therefore
worthy of consideration. What Mr. Cole has done may be repeated a
thousand times—it will be, too—when farmers have a clear view of its
attendant advantages.

How is this to be introduced to the farming world? No matter how enthusiastic and earnest the author may be in pushing it, you, who know how much hard work it requires to induce the farmer to take hold of tile drainage costing $10 or $20 per acre, can readily appreciate that they are not going to rush precipitously into a scheme requiring the expenditure of $100 to $300 per acre. But many farms are destitute of springs or permanent running waters. To such, an everflowing spring would be of great value, both for household use and for general farm purposes. One acre can be prepared at much less cost than cisterns can be dug, or wells bored, and windmills erected, aud this will secure a constant flow of perfectly pure, living water. Now if in addition to this, the farmer finds this acre more easily tilled, and yielding four-fold crops of better quality, then will he not hesitate to invest the profit of this acre upon other acres, until his farm is honeycombed with reservoirs and overflow trenches. The farmer or the gardener is then the absolute master of the situation. He cares not for floods or drouths. He knows that when he plants he shall reap an abundant harvest.

Were this system once generally adopted along the banks of the Alleghany, Monongahela, and the Ohio rivers, for a distance of only ten miles on each side, covering an area of 30,000 square miles, then would our terrible floods be reduced to insignificance, and the summer stage brought up so that the 'Broad horns' could float from Pittsburgh to Cairo in July as in March. Then would we snap our fingers at Uncle Sam, bid him take out his dams and dikes, keep hands off, and give our beautiful Ohio free swing from the Alleghany to the Mississippi. To our short-sighted vision such conception seems almost like the vaporings of a madman. But let us roll back the wheel of time for only ten years to 1876. We see tile drainage in its infancy struggling feebly for a foothold. Look upon it to-day as with giant strides it steps across the States of Ohio, Indiana, Illinois, Iowa, and Michigan, and reaches out with mighty arms to take in all of Uncle Sam's dominions, and the rest of the civilized world. To-day I have the pleasure of looking into the faces of scores of men, each one of whom could tell the wonderful story of the marvelous march of tile drainage, which if told in '76 would have marked him as a visionary and a dreamer."

The New York State Experimental Station at Geneva was among the first institutions of its kind set in operation. This station, working under a Board of Control, with Dr. Lewis E. Sturtevant as Director. In its annual report of 1888, on the first fifteen pages, is clearly set forth the fundamentals of agriculture and horticulture in fullest of perfection. Dr. Sturtevant, one of the most advanced thinkers and workers of the country, was the first to step to the front and boldly make declaration under head of "water-table" as follows: "As evaporation during the growing season is in excess of the rainfall, it becomes evident that farming success is only possible from the existence of a stored water supply within the soil from the surplus of the season, when evaporation is somewhat checked, or from the waters accumulated from the surface or saturated flow." The requirements of the principles set forth are all met by this system.

Mr. Cole's system has received the compliment of extended notice and endorsement at the hands of the leading papers of the day and country. We, in behalf of Mr. Cole, to these gentlemen extend thanks.

Since boyhood Mr. Cole has made study, observation, and experiments with a view to (by some method as near nature's as possible) save the waters of rain and snow, and hold them in storage until such time as they would be required by the crops of the growing season, to compel them to deposit their valuable elements of fertilization in the soil, and to prevent their loss in rush of floods over the earth's surface, carrying in their flow destruction to soil, crops, and other property, also prevent their stagnation, a condition in which they work injury to man, beast, and crop, breeding for the first two disease, and destruction to the last by sure and direct means. He began making earnest aud extended experiments in this line while residing in Basswood Cottage on the White Creek, Allegany County, N. Y., afterward continuing them on a plot of two city lots in the City of Brooklyn. Upon his return to his home in Wellsville, N. Y., he took up the work in earnest, and devoted time, labor, and money without stint until he had perfected his present method, and secured the desired results. Having done so, he, acting under the advice of friends who knew a little of the outlay he had been to, as well as the amount of work he had done, applied for a patent to cover the system. Never were the merits of a thing for which patent was asked more thoroughly looked into than were those of this.

On the 22d of July, 1884, the patent was allowed, and the papers issued. Underground storage and distribution of water, as well as all materials which can be used in construction are covered by it. Mr. Cole at once said a very small remuneration for his work, study and outlay was.

all he desired, and the price for right of use should be placed at a figure so low it would not be a burden or tax on the poorest landowner.

The following price has been fixed upon for right of use by individuals, and will not be varied from.

For one acre or less........................ $5 00
For all over one acre..................... 1 00 per acre.

This is but a single payment, to be made on application for deed, and not an annual payment as some have seemed to think.

Hot water rights not included in this price.

Special prices given for them on application. Mr. Cole, on his model plot grows all the most desirable varieties of strawberries, raspberries, and blackberries, currants, vines, bushes, and trees, all of which are absolutely free from disease. (See report of Dr. Early.) They are very strong, bear transportation to long distances, and take hold in new growth as soon as reset.

Special directions sent with each order sent out by us for all year round care, according to locality where sent. Orders carefully filled and packed. Prices for stock given on application. These will at all times be as low as any good stock in the market, and when the advantage of healthy growth given them by this system is considered, they will be cheaper.

For rights or any desired information, address

A. P. Cole, Agent,

Wellsville, Allegany County,

New York.

I endorse all found in this book, my son having done the work to my entire satisfaction and this is all, in my enfeebled state of health, I can contribute to its pages. As regards to the extent, value, and importance of my discoveries, I have left their vaunting to a cloud of witnesses all along uncounted, and now unanimous in their verdict of approval, and I confess that in view of the fact my discoveries are, figuratively speaking at least, that mysterious three in one and one in three of God in nature symbolized by the predictions of the earliest sages, seeking through ages to find it out. It appears at last in the form of harvests of wealth and worth to the world, transcending in value all others hitherto combined. I see no reason why I should not have at least the credit, whether living or dying, due from their coming. To say the least, it would seem settled that the water supply of the greatest city of the world, instead of being furnished from an artificial lake of unprecedented proportions, standing as a constant manace to the lives and property of millions of people, should no longer receive serious consideration, but that the counties of Westchester, Putnam, Duchess, Columbia, and others east of the Hudson, reaching out to the Adirondacks, should be dotted with living springs, linked by streams with crystal lakes, with water at temperature to grow brook trout, and in the meantime, in securing such water supply, the lands trenched will become worth on the average from fifty to one hundred dollars an acre at least, in place of one dollar in value of the past.

A. N. COLE.

Died. — At about midnight, July 14th, at his residence, "The Home on the Hillside," at Wellsville, Allegany County, N. Y., the Hon. A. N. Cole, in the 67th year of his age.

The death of Mr. Cole took place after a long illness, during which he was a great sufferer. While unable to sit up or hold his pen he dictated letters to many who will be our readers, also to the press; always anxious to lead others to see what he has known, the advantages of saving and storing one of God's best gifts to man—the waters. Since his death the press of the country of all classes, from ocean to ocean and from the lakes to the gulf, have united in paying compliment to his manhood and works. Two of these we will here lay before you. The following appeared in the *Sun* of July 16th, and is from the pen of Charles A. Dana, the ablest and most distinguished editor in the United States. The editor of the *Sun* and Mr. Cole were intimate friends:

We learn with deep regret that Asahel N. Cole, the inventor of the system of subterraneous irrigation, died at his home in Wellsville, N. Y., on Sunday evening. He was in the sixty-seventh year of his age.

A complete and practical manual of his system was finished by Mr. Cole only a few weeks ago, and we understand that it is now in the hands of the printers and will soon be published. If this system should realize even a small part of the benefits he confidently expecte from it, his name will remain immortal as one of the great benefactors of mankind.

Mr. Cole was a man of a warm, earnest, and sanguine nature, faithful to his friends, faithful to his convictions of duty, and proud of being an American. Had he lived to see his system widely developed and successfully applied, his faith in it would not have been increased; and we trust that his death may not prevent an adequate test of its value. His death leaves the world poorer by all his genius, energy, enthusiasm, and faith."

The following is from *The Husbandman* of July 24th, 1889:

"A little more than a week ago Hon. A. N. Cole died at his ' Home on the Hillside,' near Wellsville, Allegany County, N. Y. For many years he had been an important factor in public affairs. To him history will give credit as a leading organizer of the Republican party, that for twenty-four years held the Presidency of the Republic, and through a great part of that time controlled every department of the general government. Mr. Cole grew to full manhood under conditions that imposed many hardships. Unaided he procured an education that, with remarkable powers of mind, aided to make him a leader of men ; and, though the party to which he held paternal relations never rewarded him with important positions, there were times when his influence in public affairs was potent, even though not publicly recognized. Perhaps no other man in the period of his greatest activity had wider acquaintance with men who, in the critical period of internal strife and the years following, managed the destinies of the country. His keen perception and alert mind gave him fitness to direct, as he did on many occasions, the acts of men to whom power was delegated by the suffrages of their fellows.

For politics he had natural fondness, intensified by indulgence and many political triumphs, from which, as a rule, others gathered the fruits. But in late years that fondness gave way to a consuming ambition, in fact the outgrowth of a long life, devoted, in large part, to the study of economic conditions attending the development of American agriculture. He believed that farmers might augment profits greatly if they would but observe more closely the operation of natural laws, and avail of beneficent influences subject to their command. In full consonance with this idea, he began a few years ago experiments on the compact and forbidding soil of the little farm that, with a comfortable house, constituted his ' Home on the Hillside.' His purpose was to deepen and disintegrate land that in its natural state was little better than clods brought by long inaction to the greatest solidity, and he reasoned that if that unproductive land could be so improved, without great cost, to make it produce crops fully equal in quality and extent to those gathered from the rich alluvial plain in view, the fact would be of immense value to thousands of farmers whose lands were of a similar character. His

scheme comprises the most effective system of irrigation ever devised—
storing the rainfall in deep trenches extending laterally along the hill
slopes. From these he expected that water sunk through the loosened
and comminuted earth would be held within the compact walls of his
trenches, to be released only by slow percolation, whereby roots of plants
might at all seasons find abundant moisture. His success was marvelous ;
that poor land, renovated by what came to be called the ' New Agriculture,' produced crops far superior in quality and extent to any gathered
from the choicest lands of the valley. In small fruits the improvement
was so great that plain, truthful statements were often met by incredulity.
The evidence was in open view. It was credited by men of science, but
those who could profit most by full acceptance refused to see the great
improvement, hence the ' New Agriculture,' certified by fruits annually
produced with unfailing certainty, was spurned, and the story discredited
as the vagary of an old man self-deceived. No doubt this incredulity
served to deprive Mr. Cole's latter years of much happiness, if not to em-
bitter feelings naturally buoyant, and possibly to hasten his departure,
when, with his abstemious habits and natural vigor, he had reason to
look hopefully far beyond the allotted three score and ten years that he
had but just passed. It was his high ambition to leave to his fellow-men a
legacy of good to be enjoyed for all time. He did not live to see that
full appreciation for which he had intense longing, but who shall say that
his triumph will not yet be that great benefaction for which he seemed
ready to give his life ? "